SpringerBriefs in Energy

SpringerBriefs in Energy presents concise summaries of cutting-edge research and practical applications in all aspects of Energy. Featuring compact volumes of 50 to 125 pages, the series covers a range of content from professional to academic. Typical topics might include:

- A snapshot of a hot or emerging topic
- A contextual literature review
- A timely report of state-of-the art analytical techniques
- An in-depth case study
- A presentation of core concepts that students must understand in order to make independent contributions.

Briefs allow authors to present their ideas and readers to absorb them with minimal time investment.

Briefs will be published as part of Springer's eBook collection, with millions of users worldwide. In addition, Briefs will be available for individual print and electronic purchase. Briefs are characterized by fast, global electronic dissemination, standard publishing contracts, easy-to-use manuscript preparation and formatting guidelines, and expedited production schedules. We aim for publication 8–12 weeks after acceptance.

Both solicited and unsolicited manuscripts are considered for publication in this series. Briefs can also arise from the scale up of a planned chapter. Instead of simply contributing to an edited volume, the author gets an authored book with the space necessary to provide more data, fundamentals and background on the subject, methodology, future outlook, etc.

SpringerBriefs in Energy contains a distinct subseries focusing on Energy Analysis and edited by Charles Hall, State University of New York. Books for this subseries will emphasize quantitative accounting of energy use and availability, including the potential and limitations of new technologies in terms of energy returned on energy invested. The second distinct subseries connected to SpringerBriefs in Energy, entitled Computational Modeling of Energy Systems, is edited by Thomas Nagel, and Haibing Shao, Helmholtz Centre for Environmental Research - UFZ, Leipzig, Germany. This sub-series publishes titles focusing on the role that computer-aided engineering (CAE) plays in advancing various engineering sectors, particularly in the context of transforming energy systems towards renewable sources, decentralized landscapes, and smart grids.

All Springer brief titles should undergo standard single-blind peer-review to ensure high scientific quality by at least two experts in the field.

Mohammad Yaseen Mir • Javid A. Parray

Nanoenergy Production

Advances in Bioinspired Materials

Mohammad Yaseen Mir
Centre of Research for Development
University of Kashmir
Srinagar, Jammu and Kashmir, India

Javid A. Parray
Department of Environmental Science
Govt Degree College Tangdhar
Kupwara, Jammu and Kashmir, India

ISSN 2191-5520 ISSN 2191-5539 (electronic)
SpringerBriefs in Energy
ISBN 978-3-032-20858-3 ISBN 978-3-032-20859-0 (eBook)
https://doi.org/10.1007/978-3-032-20859-0

This Springer imprint is published by the registered company Springer Nature Switzerland AG
The registered company address is: Gewerbestrasse 11, 6330 Cham, Switzerland

Preface

The accelerating global demand for sustainable energy solutions, coupled with growing concerns over environmental degradation and climate change, has created an urgent need for alternative energy technologies that are efficient, scalable, and environmentally responsible. Nanoenergy production has emerged as a promising frontier in this pursuit, offering innovative pathways to harvest, convert, and store energy from ubiquitous and often underutilized sources, such as mechanical motion, heat, light, and chemical gradients. Among the many approaches explored to date, bioinspired materials and systems stand out for their ability to translate nature's time-tested strategies into advanced functional materials and devices.

Nature provides an extraordinary library of structures, materials, and mechanisms that achieve optimal performance under strict constraints of efficiency, adaptability, and sustainability. From the hierarchical architectures of human skin and insect antennae to the surface microstructures of plant leaves, fish scales, and shark skin, biological systems demonstrate how multifunctionality, robustness, and sensitivity can be achieved simultaneously. By emulating these principles, bioinspired nanoenergy devices—particularly triboelectric nanogenerators, piezoelectric systems, and biohybrid energy platforms—have shown remarkable advances in energy harvesting efficiency, self-powered sensing, and long-term operational stability.

This book, *Nanoenergy Production: Advances in Bioinspired Materials*, provides a comprehensive overview of recent progress in the design, fabrication, and application of nanoenergy systems inspired by biological and natural phenomena. It highlights how bioinspired material selection, surface morphology engineering, and structural optimization can significantly enhance device performance while enabling new functionalities such as self-powered electronic skins, neuromorphic sensors, wearable energy harvesters, and biohybrid platforms for carbon dioxide conversion. Particular emphasis is placed on triboelectric nanogenerators and related technologies, which have demonstrated exceptional versatility due to their simple architecture, wide material choices, and compatibility with flexible and stretchable substrates. Beyond performance enhancement, this book also addresses the critical challenges of sustainability and environmental impact. The use of biodegradable materials, recycled polymers, and naturally derived components is discussed as a pathway toward greener nanoenergy technologies. By integrating insights from materials science, nanotechnology, biology, and environmental engineering, the chapters collectively underscore the importance of designing nanoenergy systems that are not only high-performing but also environmentally conscious and suitable for large-scale deployment.

Srinagar, Jammu and Kashmir, India Mohammad Yaseen Mir
Kupwara, Jammu and Kashmir, India Javid A. Parray
December, 2025

Competing Interests The authors have no competing interests to declare that are relevant to the content of this manuscript.

Contents

Chapter 1
Nanomaterial–Biological Hybrid Systems for Solar-Driven CO_2 Reduction: Recent Advances and Chemical Insights

Abstract Nanomaterial biological hybrid systems (NBHSs) integrate photocatalytic nanomaterials with enzymes and microorganisms to couple efficient light harvesting with high biological selectivity. This emerging multidisciplinary approach offers significant potential for solar-to-chemical conversion, yet the fundamental interactions between abiotic nanomaterials and biotic components remain insufficiently understood. This review summarizes recent advances in material–enzyme and material–microbial hybrids with particular emphasis on electron-transfer mechanisms at nano–bio interfaces. It highlights how cytochrome and coenzyme mediated interfacial electron transport enhances synergy, enabling even non-photosensitive bacteria to perform light driven metabolic processes and CO_2 reduction. By outlining key design strategies and mechanistic insights, the chapter provides a foundation for developing next generation NBHSs that support high efficiency energy conversion and contribute to carbon neutral technologies.

Keywords Solar energy conversion · CO_2 reduction · Nanomaterial-enzyme hybrid system · Nanomaterial-microbial hybrid system

1.1 Introduction

The continued global dependence on fossil fuels has resulted in a pronounced rise in atmospheric carbon dioxide (CO_2) concentrations, which have now approached approximately 427 parts per million, as reported by the NASA Global Climate Change database. This sustained increase in atmospheric CO_2 is widely recognized as a primary driver of global warming, which in turn accelerates climate-related hazards such as soil desertification, sea-level rise, and the growing frequency and intensity of extreme weather events [1]. Current assessments further suggest that nearly 80–85% of the world's total energy supply is still derived from fossil fuel combustion, directly contributing to excessive CO_2 emissions. In light of the dual challenges of environmental deterioration and long-term energy sustainability, there is an urgent need for innovative and effective strategies to mitigate atmospheric CO_2

M. Y. Mir, J. A. Parray, *Nanoenergy Production*, SpringerBriefs in Energy,
https://doi.org/10.1007/978-3-032-20859-0_1

levels. Recent progress in carbon capture and storage (CCS) technologies has opened new pathways for addressing this issue by enabling the capture of CO_2 from industrial and atmospheric sources. When coupled with renewable energy inputs, these technologies make it increasingly feasible to convert captured CO_2 into value-added chemicals or synthetic fuels, thereby closing the carbon loop and supporting a more sustainable, low-carbon energy economy [2].

Luo et al. [3] highlights that this emerging strategy holds considerable promise for reducing the impacts of climate change and supporting the transition toward a sustainable energy landscape. A variety of environmentally responsible routes exist for converting CO_2 such as photocatalysis, electrocatalysis and biological carbon fixation. Among these, photocatalytic CO_2 reduction is particularly compelling because sunlight is abundant and freely available across most regions of the world [4]. Moreover, electrocatalytic and biocatalytic processes can also be powered by solar energy, making them compatible with renewable energy systems. Natural photosynthesis performed by plants and algae remains the dominant global mechanism for CO_2 fixation and plays an essential role in the Earth's carbon cycle. However, Blankenship et al. [5] note that its overall efficiency in converting and storing solar energy is typically below 1%, which significantly limits its potential for large scale carbon capture and fuel production. To address this limitation, nanomaterial-biological hybrid systems (NBHS) have been developed to imitate the core principles of photosynthesis while surpassing its performance. By integrating biocatalysts with engineered photosensitizers, NBHS often achieve much higher energy conversion efficiencies than natural photosynthetic pathways [6]. These hybrid systems combine biological components such as enzymes and microorganisms with artificial photocatalysts, including molecular photosensitizers and nano-semiconductors. This integration enables CO_2 conversion under mild and environmentally benign conditions, typically at room temperature and atmospheric pressure, and without requiring additional energy inputs. Furthermore, unlike plant chlorophyll, which absorbs only a narrow portion of the solar spectrum, semiconductor materials in NBHS have broad light absorption capabilities. When layered or combined strategically, these materials can provide complementary absorption across multiple regions of the solar spectrum [7]. Such enhanced light utilization significantly improves the efficiency of solar energy harvesting and boosts the production of organic compounds generated through CO_2 reduction [8]. Overall, the present work concentrates on the design, optimization, and application of NBHS as an advanced platform for sustainable CO_2 conversion. Although NBHS possess a far simpler architecture than natural photosynthetic machinery and offer clear advantages such as modularity, replaceability, and upgradability, their performance is often constrained by limited biocompatibility at the abiotic–biotic interface. This interfacial incompatibility hampers efficient charge transfer, thereby making it difficult to achieve high catalytic efficiencies and precise chemical selectivity [9]. Consequently, the development of flexible and highly integrated NBHS has emerged as a key research priority. An effective NBHS must address several interrelated challenges, most notably the enhancement of interfacial electron transport and the creation of surfaces that are inherently biocompatible with biological catalysts. In response, recent studies have concentrated on optimizing both the photosensitizers and biocatalysts employed in bio–nano hybrid systems, with particular emphasis on tailoring their electronic structures and interfacial interactions [10]. To mitigate the limitations associated with inefficient electron transfer across the bio–inorganic interface, a range of innovative strategies has been proposed, including surface functionalization, nanostructure engineering, and the use of redox mediators to facilitate biocompatible photocatalysis and biocatalysis. Table 1.1 provides a comparative overview of representative NBHS designs, summarizing their nanostructure types, corresponding biological components, underlying reaction mechanisms, and key performance metrics. This comparison highlights current progress as well as remaining challenges in the rational design of high-performance, biocompatible NBHS.

1.2 Nanomaterial-Enzyme Hybrid System

When enzymes are integrated with photocatalytic nanomaterials, photoelectrons can be transferred with much higher efficiency, enabling rapid conversion rates and nearly 100% selectivity toward targeted organic products. This combined strategy effectively bypasses the kinetic and thermodynamic limitations that normally hinder the production of such compounds [15]. However, constructing these hybrids systems referred to as enzymatic NBHS is highly complex because it requires precise and stable interactions between biological enzymes and inorganic surfaces. As a result, researchers have explored multiple enzyme immobilization techniques on nanomaterials, including covalent attachment and electrostatic adsorption. These approaches improve the electron flow from the nanomaterial to the enzyme's redox-active cofactor, thereby enhancing catalytic activity. Recent progress in enzymatic NBHS has also introduced more sophisticated configurations, such as electrode-assisted photoelectrochemical hybrid systems. A key factor governing the efficiency of CO_2 bio-reduction is the rate of electron transfer, as nanomaterials must reliably deliver photoexcited, high-energy electrons to the catalytic enzyme. This transfer can occur through two main mechanisms: direct or indirect pathways. In direct electron transfer (DET), photoelectrons move straight from the nanomaterial to the enzyme's internal redox center such as heme groups or flavin cofactors (Fig. 1.1). Achieving efficient DET requires nanomaterials with suitable dimensions, biocompatibility, and surface chemistry to ensure strong enzyme–nanomaterial coupling [16]. In contrast, indirect electron transfer (IET) relies on redox mediators including methyl viologen, neutral red, and various Rh complexes which shuttle photoelectrons from the photocatalyst to the enzyme's active site.

1.2.1 Direct Electron Transfer Drives Enzymatic Reactions

Several enzymes such as ferredoxin reductases, formate dehydrogenase (FDH), and carbon monoxide dehydrogenases (CODH) naturally drive CO_2-reduction pathways [17]. Their catalytic activity relies on [Fe_4S_4] clusters, which readily accept and

Table 1.1 Representative nanomaterial–biological hybrid systems for solar-driven CO_2 conversion

Nanomaterial component	Biological component	Mechanism	Target products	Light source	Performance highlights	Reference
CdS nanoparticles	*MoFe* nitrogenase	Photo-induced electron transfer to enzyme	Ammonia	Visible light	Enhanced N_2 and CO_2 reduction under light	[8]
TiO_2 nanoparticles	*Rhodopseudomonas palustris*	Photocatalytic activation + microbial fixation	Acetate	Simulated sunlight	CO_2-to-acetate conversion with 85% selectivity	[11]
Graphene oxide	*Synechocystis* sp. PCC 6803	Enhanced light absorption; photosynthetic metabolism	Lactate, ethanol	Natural sunlight	1.8× yield improvement in organic acids	[12]
Cadmium-based QDs/CdTe interfaced with autotrophic bacteria (quantum-dot–bacteria hybrids)	*Xanthobacter autotrophicus*, *Rhodopseudomonas palustris*	Quantum dots harvest visible/near-IR light and transfer electrons to microbial CO_2 fixation pathways	Value-added organics, biomass, CH_4	Sunlight	Broad spectral absorption, tunable bandgaps, demonstrated increases in light-driven fixation rates	[13]
NiCu-decorated CdS on *Methanosarcina barkeri* (assembled hybrid)	*Methanosarcina barkeri* (methanogen)	Photoexcited electrons from NiCu@CdS delivered to methanogenic pathway	CH_4 (high selectivity)	Solar driven	Very high CH_4 selectivity and improved quantum yield (proof-of-concept for solar biomethanation)	[14]

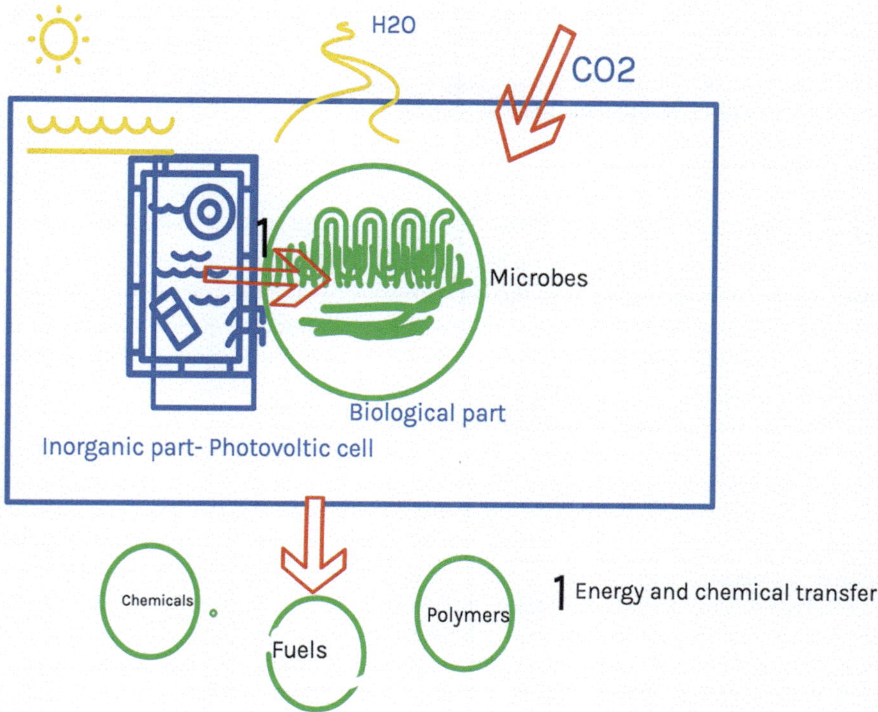

Fig. 1.1 Solar-driven biological inorganic hybrid systems for the production of solar chemicals and fuels from carbon dioxide

shuttle electrons to active sites. For abiotic photocatalysts, designing an advanced nano-architectural framework that can effectively host these [Fe_4S_4] clusters or enzyme catalytic domains is therefore essential. Such architectures promote efficient transfer of photoexcited electrons, enabling their participation in enzyme-regulated redox reactions. This underscores the importance of biomimetic nanostructural design to enhance photocatalytic performance by replicating biological electron transfer mechanisms [18]. This electron transfer route operates without dissociative redox mediators and is instead governed by direct electron transfer (DET). Among enzyme nanomaterial hybrids, hydrogenases are widely studied due to their internal [Fe_4S_4] clusters for electron relay and iron-based metal centres that catalyse proton reduction. Inspired by representative hydrogenase based NBHS designs, numerous nanomaterials such as CdS, CdTe, carbon dots, and TiO_2 have been explored as photocatalysts capable of generating excited electrons and injecting them into enzymatic active sites for CO_2 reduction [19]. For example, formate dehydrogenase from *Desulfovibrio Hildenborough* (DvFDH) has been immobilized on hollow-glass-microsphere/TiO_2 nanoparticle (HGM/TiO_2 NP) systems, where the photosensitizer supplies photoelectrons to drive CO_2-to-formate conversion. These enzyme nanomaterial assemblies exemplify DET based catalysis, as each enzyme can internally shuttle electrons through its protein matrix to promote the reaction. To counteract recombination effects, the holes generated alongside photoelectrons must be neutralized using electron donors such as EDTA, water, triethanolamine (TEOA), or ascorbic acid (AA) [20].

1.2.2 Indirect Electron Transfer Drives Enzymatic Reaction

Many native enzymes such as glutamate dehydrogenase, NADPH-dependent alcohol dehydrogenase, and old yellow enzyme cannot directly draw electrons from solid surfaces and instead rely on intermediate electron-transfer steps [21]. In natural photosynthesis, chlorophyll generates photoelectrons that are relayed along the electron-transport chain via ferredoxin, which acts as a mediator to regenerate the NADH cofactor [22]. Likewise, several nanomaterial–enzyme hybrid systems depend on redox mediators (M), such as methyl viologen (MV^{2+}), to shuttle photoexcited electrons from the photocatalyst to the enzyme [23]. In these mediated electron transfer (MET) systems, the enzyme accepts electrons in a sequential manner to drive catalysis, while the semiconductor photocatalyst continually provides electrons during illumination. NADH a readily accessible and stereospecific cofactor often serves as the mediator in milder NBHS configurations, delivering protons and electrons to the enzymatic active site [11]. Using NADH simplifies photocatalytic carbon-sequestration designs because it circumvents the need for complex biocompatible interfaces that support direct electron exchange. Importantly, IET-assisted NBHS platforms significantly enhance both NADH stability and regeneration efficiency [24]. For example, diatom-inspired graphite-carbon nitride (g-C_3N_4) can harvest solar energy to regenerate NADH through interaction with Rh-based electron mediators, achieving regeneration efficiencies above 100% after 2 h of reaction. Such mediator–photocatalyst pairs are routinely combined with NADH-dependent FDH to convert CO_2 into formate under solar illumination. Recent studies reported a bromophenol-phenolic (BPB) composite that exhibits greatly enhanced photocatalytic performance [25]. In BPB–enzyme hybrid systems, visible light absorption triggers electron-transfer events that regenerate NADH via a Rh mediator, enabling CO_2-to-formate conversion by FDH. The BPB composite achieves an NADH regeneration efficiency of 82.52%, far surpassing conventional photocatalysts such as colloidal TiO_2, doped TiO_2, or CdS [14]. Artificial-chloroplast research has introduced additional strategies for CO_2 reduction. For instance, CdS-coated arginine titanium orange (PTi) microcapsules markedly improved NADH regeneration due to favourable electron alignment and efficient charge separation. This system achieved electron–hole separation efficiencies of 93.03 ± 3.84% and NADH regeneration rates of 4226 ± 121 $\mu mol\ g^{-1}\ h^{-1}$. Using individual or coupled enzymes, the bioinspired chloroplast architecture produced formate and methanol at rates of 1500 and 99 $\mu M\ h^{-1}$, respectively. Furthermore, a novel photosensitizer was created by conjugating Rh complexes with benzodiazepine-based polymeric carbon nitride (PCN), enabling highly selective conversion of NAD^+ to 1,4-NADH. This strategy broadened the utility of diverse electron mediators for enzymatic catalysis and provided a practical approach for immobilizing mediators to support continuous, light-driven CO_2-to-formate conversion in NBHS platforms [26].

1.3 Nanomaterial-Microbial Hybrid System

Another way to produce solar chemicals is to build a microbial photoelectrochemical system that combines microorganisms with photoelectrodes. When exposed to sunlight, the photoelectrodes generate two electrons and a hole. The holes are recombined by the electrons from the anode in the photocathode or by the sacrificial agents (such water and organics) in the photoanode. Meanwhile, microorganisms suspended in the medium or adhered to the cathode surface utilize electrons or electron mediators such as H_2 to convert CO_2 into valuable products. Because photovoltaic cells are not part of the external circuit, microbial photoelectrochemical systems have a simpler configuration compared with photovoltaic-powered bio–inorganic hybrid setups. Additionally, photoelectrodes integrate both light harvesting and electrochemical catalysis into a single component, whereas photovoltaic-driven hybrid systems require separate units to perform these two functions. Nichols et al. [27] reported a bio-inorganic hybrid microbial photoelectrochemical platform for converting solar energy into chemical fuels. The system used a biocompatible n^+/p-Si photocathode coated with a Ni–Mo alloy to generate H_2, which served as the electron donor enabling *Methanosarcina barkeri* (M. barkeri) to reduce CO_2 into CH_4. This configuration achieved methane production with a Faradaic efficiency of 82 ± 10% at a low overpotential of 175 mV. In related work, a fully self-powered solar-to-methane system was demonstrated by replacing the n^+/p-Si electrode with a p-InP photocathode overlaid with platinum [27]. However, using a single semiconductor material as the photoelectrode often results in limited electron output and fast recombination of photogenerated charge carriers. To overcome these issues, a $WO_3/MoO_3/g\text{-}C_3N_4$ heterojunction photocathode was recently introduced, offering improved photoelectrochemical performance in microbial systems [28]. The formation of a heterojunction can effectively enhance redox potentials while promoting the efficient separation of photogenerated charge carriers. In previously reported microbial photoelectrochemical systems, the H_2 generated at the photocathode served as an electron mediator, enabling microorganisms to reduce CO_2 into value-added chemicals. However, the use of H_2 as an intermediate suffers from two major drawbacks: its low solubility in aqueous electrolytes and the high overpotential required for hydrogen evolution [29]. Owing to its limited solubility, the local concentration of H_2 near the microorganisms may be insufficient, thereby restricting the microbial reduction of CO_2. To overcome these limitations, microbial photoelectrochemical systems should ideally avoid reliance on the hydrogen evolution reaction. In this context, Fu et al. recently proposed an innovative microbial photoelectrochemical system that integrates a TiO_2 photoanode for solar energy absorption with a biocathode capable of directly converting CO_2 into CH_4 without the involvement of H_2 as an electron mediator. Upon light illumination, the TiO_2 photoanode generates electron–hole pairs; the photogenerated electrons are then transported through an external circuit to the biocathode, where they directly drive microbial CO_2 reduction. This direct electron transfer pathway not only bypasses the drawbacks associated with H_2 mediation but also improves the overall efficiency and selectivity of CO_2 conversion. Crucially, methanogens that form biofilms on the cathode surface can directly draw electrons from the electrode to reduce CO_2, eliminating the need for H_2 as an intermediate. This direct electron transfer (DET) between the cathode and microbes enabled the biocathode to operate at an exceptionally low overpotential below 50 mV for CH_4 production. Coupling a TiO_2 photoanode with this DET enabled biocathode produced CH_4 with outstanding stability and selectivity, achieving a Faradaic efficiency of up to 96% over 90 h. At the time, this FE represented the highest reported value in CO_2 electrochemical or photoelectrochemical reduction. Expanding the system's light-harvesting capability, Xiao et al. [7] replaced the TiO_2 photoanode with a TiO_2/CdS heterostructure, broadening absorption from the UV region (<400 nm) into the visible range (<500 nm) [7]. As a result, the solar-to-chemical efficiency (SCE) increased fivefold compared to the TiO_2-only system, demonstrating that enhanced light capture is an effective strategy for improving SCE. The optimized hybrid system achieved a peak SCE of 1.28% at an illumination intensity of 25 mW cm^{-2} about five times higher than the global average efficiency of natural photosynthesis. Liu et al. [30] later introduced an unassisted microbial photoelectrochemical platform inspired by the natural photosynthesis. This setup used a Si nanowire photocathode and a TiO_2 nanowire photoanode, with *S. ovata* immobilized on the Si nanowires as the microbial catalyst. By integrating two semiconductor photoelectrodes, the system effectively used the full solar spectrum. It produced acetic acid with a high Faradaic efficiency of 90% and an overpotential below 200 mV over 20 h. The acetic acid could then be upgraded by engineered *E. coli* into higher-value compounds such as n-butanol and polyhydroxybutyrate. Owing to the dual-photoelectrode design, the system achieved an SCE of 0.38%, significantly higher than the 0.1% obtained using only a TiO_2 photoanode [30]. Notably, the structured Si nanowire photocathode also created a localized anaerobic microenvironment, allowing CO_2 reduction even under ambient air containing 21% O_2.

In microbial photoelectrochemical systems, CO_2 reduction at the cathode primarily depends on electron transfer between the electrode and the bacteria. This electron flow can occur either through direct electron transfer (DET) from the cathode to the cells or via an indirect electron transfer (IDET) pathway, where H_2 functions as the key electron mediator. A major challenge for these systems is improving electron-transfer efficiency reflected in the Faradaic efficiency, which measures how effectively electrons in the circuit are converted into chemical products. Additional limitations include low microbial biomass on the cathode and slow bacterial attachment to its surface. These issues stem largely from the inherently slow growth rates of the microorganisms and their weak interactions with the electrode material [31], both of which restrict solar-to-chemical efficiency (SCE) and product formation rates. Future work will therefore aim to enhance bacterial proliferation, strengthen microbe electrode interactions, and accelerate electron transport from the cathode to the microorganisms.

1.4 Charge Transfer Between Microorganisms and Inorganic Materials

Solar powered bio–inorganic hybrid systems integrate light-absorbing materials for photon capture with microbial catalysts that carry out CO_2 reduction. In these systems, microbes rely on electrons or reducing equivalents most commonly H_2 to drive their CO_2 conversion pathways. The efficiency of the entire hybrid platform hinges on the charge-transfer route and rate between the microorganisms and the inorganic components. However, the exact mechanisms governing electron exchange between microbes and inorganic materials, such as electrodes or nanoparticles, remain poorly understood.

To develop effective bio–inorganic hybrids, it is essential to use inorganic materials that exhibit appropriate surface properties and strong biocompatibility. Gaining deeper insight into the microbial–material interface is therefore critical, as it will enable the rational design of new hybrid architectures capable of converting solar energy into chemical products. In microbial photoelectrochemical systems and photovoltaic-driven hybrids, charge transfer occurs between the microorganisms and the electrode, whereas in photosensitized bio–inorganic systems, electron exchange takes place between microbes and photocatalyst nanoparticles. We also describe strategies to improve the interaction between microorganisms and inorganic materials, such as the transfer of charge between them and the adherence of bacteria to inorganic materials. DET and IDET are the two primary channels via which electrons move from the cathode to the microbes. A number of exterior membranes (Direct contact) and periplasmic complexes, including pili on the surface of microorganisms (Pili contact), are used to carry out the DET pathway [32]. Electron mediators, namely H_2, are used to carry out the IDET pathway. However, H_2 has two disadvantages as an electron mediator: it is poorly soluble and has a large potential for producing H_2 [32]. As a result, the DET route is considered the more efficient mode of electron exchange between microbes and electrodes. In microbial photoelectrochemical systems, microorganisms adhered to the cathode can directly take up electrons to reduce CO_2 [7]. When an external bias from a solar cell or potentiostat is applied, the cathode may also produce H_2 at high overpotentials, which microbes can then use as an electron mediator for CO_2 reduction [28]. However, the limited solubility and slow diffusion of H_2 in the electrolyte lead to energy losses and reduced electron-transfer efficiency.

Because DET consistently achieves higher Faradaic efficiencies for generating organic products (>90%) compared to IDET (<90%), the chosen electron-transfer pathway largely determines the overall FE of hybrid systems. Thus, DET offers a more effective and promising electron-delivery mechanism. Optimizing operational conditions such as minimizing overpotential to prevent H_2 formation can further decrease electron-transfer resistance and enhance efficiency.

In photosensitized biological–inorganic hybrid systems, microorganisms are directly coupled with photocatalyst nanoparticles, enabling intimate interfacial charge transfer. Upon light irradiation, photogenerated electrons from the photocatalysts can be directly accepted by membrane-bound hydrogenases or other redox-active membrane proteins, such as cytochromes, on the microbial surface [33]. In addition to direct electron transfer, indirect pathways involving electron mediators—such as H_2 produced by the photocatalysts or soluble redox shuttles—may also participate. However, some of these mediators can be consumed in competing reactions or diffuse into the bulk electrolyte, thereby reducing charge utilization efficiency. Compared with microbe–cathode configurations, microorganism–photocatalyst systems typically form well-dispersed colloidal suspensions, which provide an excellent platform for probing charge transfer mechanisms using advanced spectroscopic techniques. For example, Kornienko et al. investigated the electron transfer kinetics between CdS photocatalyst nanoparticles and *Moorella thermoacetica* by combining transient absorption (TA) and time-resolved infrared (TRIR) spectroscopy with biological activity measurements [34]. Their results revealed two distinct electron transfer pathways between the photocatalyst and the microorganism, governed by the presence or absence of hydrogenase. When hydrogenase was absent, electron transfer from the photocatalyst to the microorganism was inefficient during CO_2 reduction, primarily due to slow charge transfer kinetics that promoted rapid recombination of photogenerated electrons and holes. In contrast, the presence of membrane-bound hydrogenase significantly enhanced electron transfer efficiency, as its fast kinetics facilitated rapid electron extraction from the photocatalyst and suppressed recombination [35]. This enhancement enabled effective transport of electrons from the microbial surface into the cytoplasm, where they were utilized for CO_2 reduction. Overall, these findings establish a clear correlation between hydrogenase activity and CO_2 reduction performance in photosensitized biological–inorganic systems. They highlight the critical role of membrane-bound hydrogenase in mediating efficient interfacial electron transfer and underscore its importance for improving the catalytic efficiency of microbial CO_2 conversion. Extensive research indicates that engineering nanostructured surfaces on inorganic materials is an effective strategy for enhancing microbe–material interactions. Their high surface area, mechanical robustness, and dimensions comparable to microbial cells make these nanostructures particularly advantageous [35]. Motivated by this concept, Liu et al. developed a microbial photoelectrochemical platform for converting CO_2 to acetate by integrating *S. ovata* with Si nanowire arrays. The Si nanowires exhibited strong biocompatibility and offered abundant attachment sites for microbial colonization. In addition, the nanowire architecture improved charge separation and reduced electron–hole recombination, while maintaining adequate spacing to ensure unhindered mass transport within the array [36]. It was also observed that microorganisms such as *S. oneidensis* MR-1 can recognize and preferentially align along Si nanowire structures, forming well-organized biofilms.

Conductive nanomaterials—such as graphene and carbon nanotubes—can be incorporated during biofilm development to enhance its conductivity, create additional electron-transport pathways, and further reduce barriers to electron transfer. Surface chemical modifications of inorganic materials, through mechanisms like hydrogen bonding, van der Waals forces, and electrostatic interactions, can also significantly improve microbial adhesion and charge-transfer efficiency [37]. Beyond structural and surface chemistry considerations, the intrinsic biocompatibility of inorganic materials remains a major challenge in hybrid systems. While materials like CdS, InP, and NiMoZn alloys have been successfully paired with microbes, the release of toxic ions from corrosion or the formation of reactive oxygen species during photocatalysis can harm living cells. This underscores the need for environmentally benign, microbe-friendly inorganic components. In parallel, microorganisms themselves can be protected using coatings such as metal–organic frameworks (MOFs), hydrogels, or SiO_2 layers to enhance their tolerance against ROS and oxygen exposure [38].

1.5 Challenges and Opportunities

Solar-driven biological–inorganic hybrid systems hold significant promise for converting CO_2 into valuable organic molecules using sunlight. Performance is typically assessed using three key parameters: durability, Faradaic efficiency (FE), and solar-to-chemicals conversion efficiency (SCE). SCE represents the ratio of the chemical energy stored in the generated fuels to the effective solar energy input. FE reflects the proportion of electrons that contribute to fuel formation relative to the total electrons passing through the circuit. Durability describes the operational stability and lifespan of the hybrid system. Recent advances have led to notable improvements in all three metrics. In particular, many biological–inorganic hybrid systems now achieve nearly 100% FE for CO_2 reduction, significantly outperforming artificial photosynthetic systems, which often reach only ~70% FE. This superior selectivity arises primarily from the inherent biochemical specificity of microbial catalysts compared with conventional chemical catalysts. Moreover, the self-replicating nature of living microorganisms endows biological–inorganic hybrid systems with inherently greater durability and longer operational lifetimes than purely artificial photosynthetic systems. In many cases, these hybrid platforms also exhibit higher solar-to-chemical conversion efficiencies (SCEs), as biological catalysts can achieve remarkable selectivity and efficiency under mild conditions. Nevertheless, despite their advantages in terms of Faradaic efficiency (FE), SCE, and longevity, solar-driven biological–inorganic hybrid systems remain at an early stage of development. At present, both their SCE and long-term stability still fall short of the requirements for practical and industrial deployment. The primary factors limiting the efficiency and durability of these hybrid systems include inefficient charge transfer across the biotic–abiotic interface, the intrinsically slow biological response and metabolic rates of microorganisms, and the limited solar absorption capacity of current light-harvesting materials. Addressing these challenges requires the development of light absorbers with broader solar spectral coverage and higher photoconversion efficiencies. Importantly, such inorganic components including photocatalysts, photoelectrodes, and photovoltaic devices should be composed of earth-abundant, low-cost materials to ensure scalability and economic viability. In parallel, advances in synthetic biology and genetic engineering offer powerful strategies to enhance microbial performance by redesigning metabolic pathways for faster and more efficient CO_2 conversion. Engineering microorganisms with optimized redox enzymes, improved electron uptake pathways, and higher tolerance to interfacial stresses could substantially accelerate biochemical reaction rates. Additionally, tailoring the surface chemistry and micro−/nanostructure of inorganic materials can promote more effective interfacial charge transfer between the microorganisms and the abiotic components. Consequently, systematic investigation into controlling and optimizing charge transfer at the biotic–abiotic interface represents a particularly promising and worthwhile direction for future research.

References

1. Ješić D, Lašič Jurković D, Pohar A, Suhadolnik L, Likozar B (2021) Engineering photocatalytic and photoelectrocatalytic CO_2 reduction reactions: mechanisms, intrinsic kinetics, mass transfer resistances, reactors and multi-scale modelling simulations. Chem Eng J 407:126799
2. White JL, Baruch MF, Pander JE III, Hu Y, Fortmeyer IC, Park JE, Zhang T, Liao K, Gu J, Yan Y, Shaw TW, Abelev E, Bocarsly AB (2015) Light-driven heterogeneous reduction of carbon dioxide: photocatalysts and photoelectrodes. Chem Rev 115:12888–12935
3. Luo W, Xie W, Mutschler R, Oveisi E, De Gregorio GL, Buonsanti R, Züttel A (2018) Selective and stable electroreduction of CO_2. ACS Catal 8:6571–6581
4. Mitali J, Dhinakaran S, Mohamad AA (2022) Energy storage systems: a review. Energy Storage Sav 1:166–216
5. Blankenship RE, Tiede DM, Barber J, Brudvig GW, Fleming G, Ghirardi M, Gunner MR, Junge W, Kramer DM, Melis A, Moore TA, Moser CC, Nocera DG, Nozik AJ, Ort DR, Parson WW, Prince RC, Sayre RT (2011) Comparing photosynthetic and photovoltaic efficiencies and recognizing the potential for improvement. Science 332:805–809
6. Zuo WL, Yu YD, Huang H (2021) Making waves: microbe-photocatalyst hybrids may provide new opportunities for treating heavy metal polluted wastewater. Water Res 195:116984
7. Xiao KM, Liang J, Wang XY, Hou TF, Ren XN, Yin PQ et al (2022) Panoramic insights into semi-artificial photosynthesis: origin, development, and future perspective. Energ Environ Sci 15:529–549
8. Brown KA, Harris DF, Wilker MB, Rasmussen A, Khadka N, Hamby H, Keable S, Dukovic G, Peters JW, Seefeldt LC, King PW (2016) Light-driven dinitrogen reduction catalyzed by a CdS:nitrogenase MoFe protein biohybrid. Science 352(6284):448–450. https://doi.org/10.1126/science.aaf2091
9. Fang Z, Zhou J, Zhou X, Koffas MAG (2021) Abiotic-biotic hybrid for CO_2 biomethanation: from electrochemical to photochemical process. Sci Total Environ 791:148288. https://doi.org/10.1016/j.scitotenv.2021.148288
10. Hu LW, Yang JZ, Xia Q, Zhang J, Zhao HX, Lu Y (2024) Chemico-biological conversion of carbon dioxide. J Energy Chem 89:371–387
11. Chen H, Huang Y, Sha C, Moradian JM, Yong YC, Fang Z (2023) Enzymatic carbon dioxide to formate: mechanisms, challenges and opportunities. Renew Sustain Energy Rev 178:113271

12. Wang B, Jiang Z, Yu JC, Wang J, Wong PK (2019) Enhanced CO_2 reduction and valuable C_2^+ chemical production by a CdS-photosynthetic hybrid system. Nanoscale 11:9296–9301
13. Guan X, Erşan S, Hu X, Atallah TL, Xie Y, Lu S, Cao B, Sun J, Wu K, Huang Y, Duan X, Caram JR, Yu Y, Park JO, Liu C (2022) Maximizing light-driven CO_2 and N_2 fixation efficiency in quantum dot-bacteria hybrids. Nat Catal 5(11):1019–1029. https://doi.org/10.1038/s41929-022-00867-3
14. Ye J, Wang C, Gao C, Fu T, Yang C, Ren G, Lü J, Zhou S, Xiong Y (2022) Solar-driven methanogenesis with ultrahigh selectivity by turning down H_2 production at biotic-abiotic interface. Nat Commun 13(1):6612. https://doi.org/10.1038/s41467-022-34423-1
15. Yadav RK, Oh GH, Park NJ, Kumar A, Kong KJ, Baeg JO (2014) Highly selective solar-driven methanol from CO2 by a photocatalyst/biocatalyst integrated system. J Am Chem Soc 136(48):16728–16731. https://doi.org/10.1021/ja509650r
16. Patil PD, Nadar SS, Marghade DT (2021) Photo-enzymatic green synthesis: the potential of combining photo-catalysis and enzymes. In: Inamuddin, Boddula R, Ahamed MI, Khan A (eds) Advances in green synthesis: avenues and sustainability. Springer, Cham, pp 173–189
17. Tian Y, Zhou Y, Zong Y, Li J, Yang N, Zhang M, Guo Z, Song H (2020) Construction of functionally compartmental inorganic photocatalyst-enzyme system via imitating chloroplast for efficient Photoreduction of CO_2 to formic acid. ACS Appl Mater Interfaces 12(31):34795–34805. https://doi.org/10.1021/acsami.0c06684
18. Wang ZF, Hu Y, Zhang SP, Sun Y (2022) Artificial photosynthesis systems for solar energy conversion and storage: platforms and their realities. Chem Soc Rev 51:6704–6737
19. Holá K, Pavliuk MV, Németh B, Huang P, Zdražil L, Land H, Berggren G, Tian HN (2020) Carbon dots and [FeFe] hydrogenase biohybrid assemblies for efficient light-driven hydrogen evolution. ACS Catal 10:9943–9952
20. Lam E, Miller M, Linley S, Manuel RR, Pereira IAC, Reisner E (2023) Comproportionation of CO_2 and cellulose to Formate using a floating semiconductor-enzyme photoreforming catalyst. Angew Chem Int Ed Engl 62(20):e202215894. https://doi.org/10.1002/anie.202215894
21. Kim J, Lee SH, Tieves F, Choi DS, Hollmann F, Paul CE, Park CB (2018) Biocatalytic C=C bond reduction through carbon Nanodot-sensitized regeneration of NADH analogues. Angew Chem Int Ed Engl 57(42):13825–13828. https://doi.org/10.1002/anie.201804409
22. Zhang Y, Zhao Y, Li R, Liu J (2021) Bioinspired NADH regeneration based on conjugated photocatalytic systems. Solar RRL 5(2):2000339
23. Ji XY, Liu CC, Wang J, Su ZG, Ma GH, Zhang SP (2017) Integration of functionalized two-dimensional TaS_2 nanosheets and an electron mediator for more efficient biocatalyzed artificial photosynthesis. J Mater Chem A 5:5511–5522
24. Zhou JH, Yu SS, Kang HL, He R, Ning YX, Yu YY, Wang M, Chen BQ (2020) Construction of multi-enzyme cascade biomimetic carbon sequestration system based on photocatalytic coenzyme NADH regeneration. Renew Energy 156:107–116
25. Kumar S, Yadav RK, Gupta S, Choi SY, Kim TW (2023) A spherical photocatalyst to emulate natural photosynthesis for the production of formic acid from CO_2. J Photochem Photobiol A Chem 438:114545
26. Zhang PY, Dong WJ, Zhang YY, Zhao LN, Yuan HL, Wang CJ, Wang WS, Wang HX, Zhang HY, Liu J (2023) Metalated carbon nitride with facilitated electron transfer pathway for selective NADH regeneration and photoenzyme-coupled CO reduction. Chin J Catal 54:188–198
27. Nichols EM, Gallagher JJ, Liu C, Su Y, Resasco J, Yu Y, Sun Y, Yang P, Chang MC, Chang CJ (2015) Hybrid bioinorganic approach to solar-to-chemical conversion. Proc Natl Acad Sci U S A 112(37):11461–11466. https://doi.org/10.1073/pnas.1508075112
28. Cai Z, Huang L, Quan X, Zhao Z, Shi Y, Li Puma G (2020) Acetate production from inorganic carbon (HCO_3^-) in photo-assisted biocathode microbial electrosynthesis systems using WO_3/MoO_3/g-C_3N_4 heterojunctions and *Serratia marcescens* species. Appl Catal B 267:118611
29. Karthikeyan R, Singh R, Bose A (2019) Microbial electron uptake in microbial electrosynthesis: a mini-review. J Ind Microbiol Biotechnol 46(9–10):1419–1426. https://doi.org/10.1007/s10295-019-02166-6

30. Liu C, Colon BC, Ziesack M, Silver PA, Nocera DG (2016) Water splitting–biosynthetic system with CO_2 reduction efficiencies exceeding photosynthesis. Science 352:1210
31. Ploux L, Ponche A, Anselme K (2010) Bacteria/material interfaces: role of the material and cell wall properties. J Adhes Sci Technol 24(13–14):2165–2201. https://doi.org/10.1163/016942410X511079
32. Rabaey K, Rozendal RA (2010) Microbial electrosynthesis—revisiting the electrical route for microbial production. Nat Rev Microbiol 8(10):706–716. https://doi.org/10.1038/nrmicro2422
33. Sakimoto KK, Wong AB, Yang PD (2016) Self-photosensitization of non photosynthetic bacteria for solar-to-chemical production. Science 351:74–77
34. Kornienko N, Zhang JZ, Sakimoto KK, Yang P, Reisner E (2018) Interfacing nature's catalytic machinery with synthetic materials for semi-artificial photosynthesis. Nat Nanotechnol 13(10):890–899. https://doi.org/10.1038/s41565-018-0251-7
35. Xu M, Tremblay PL, Ding R, Xiao J, Wang J, Kang Y, Zhang T (2021) Photo-augmented PHB production from CO_2 or fructose by Cupriavidus necator and shape-optimized CdS nanorods. Sci Total Environ 753:142050. https://doi.org/10.1016/j.scitotenv.2020.142050
36. Sahoo PC, Pant D, Kumar M, Puri SK, Ramakumar SSV (2020) Material-microbe interfaces for solar-driven CO_2 bioelectrosynthesis. Trends Biotechnol 38(11):1245–1261. https://doi.org/10.1016/j.tibtech.2020.03.008
37. Aryal N, Ammam F, Patil SA, Pant D (2017) An overview of cathode materials for microbial electrosynthesis of chemicals from carbon dioxide. Green Chem 19:5748–5760
38. Yang SH, Lee KB, Kong B, Kim JH, Kim HS, Choi IS (2009) Biomimetic encapsulation of individual cells with silica. Angew Chem Int Ed Engl 48(48):9160–9163. https://doi.org/10.1002/anie.200903010

Chapter 2
Biomimetic Approaches in Triboelectric Nanogenerators: Design, Principles and Emerging Applications

Abstract Triboelectric nanogenerators (TENGs) have emerged as a promising energy-harvesting technology, attracting considerable attention due to their vast potential in sustainable power generation and self-powered sensing systems. The electrical output and mechanical durability of TENGs are strongly influenced by key factors such as device architecture, surface micro–/nanostructure, and the choice of triboelectric materials. Extensive research efforts have therefore focused on optimizing these parameters to enhance device efficiency and reliability. Notably, significant performance improvements in TENGs have been achieved through bio-inspired design strategies, which emulate natural structures, surface textures, material functionalities, and biological energy conversion or sensing mechanisms. By mimicking these naturally optimized systems, TENGs can achieve higher charge density, improved contact efficiency, and enhanced adaptability. Furthermore, the simple configuration, self-powered operation, and tunable electrical output of TENGs have enabled their integration into a broad range of biomimetic applications, including wearable electronics, artificial skins, and environmental sensing platforms. This chapter presents a comprehensive analysis of recent advances in bioinspired TENGs and TENG-enabled biomimetic systems. It begins with an overview of nature-inspired TENG designs reported in the literature, followed by a comparative discussion of different device structures, surface morphologies, and triboelectric materials derived from biological inspiration, along with the resulting performance enhancements. In addition, various natural power generation and ubiquitous sensing mechanisms are examined and correlated with their artificial TENG counterparts, highlighting the effectiveness of bioinspired approaches in advancing next-generation energy harvesting and sensing technologies.

Keywords Triboelectric nanogenerator · Biomimetic · Plant-inspired structural design · Animal-inspired structural design · TENG enabled electronic skin

M. Y. Mir, J. A. Parray, *Nanoenergy Production*, SpringerBriefs in Energy,
https://doi.org/10.1007/978-3-032-20859-0_2

2.1 Introduction

Since their introduction, triboelectric nanogenerators (TENGs) have attracted substantial scientific attention and have rapidly evolved into a prominent research focus owing to their strong potential in energy harvesting and self-powered sensing applications [1]. TENGs function through the coupled effects of contact electrification and electrostatic induction, enabling the direct conversion of mechanical energy into electrical energy. Their key advantages including simple structural design, broad material compatibility, high output performance, ease of fabrication, and feasibility for flexible and stretchable configurations allow TENGs to efficiently harvest ambient mechanical energy from diverse sources such as vibrations, wind, water waves, and human motion [2]. A major advantage of TENG over conventional energy harvesting technologies, including electromagnetic, piezoelectric, and photovoltaic systems, lies in their exceptional capability to scavenge irregular, low-frequency mechanical energy commonly present in the environment [3]. Beyond energy harvesting, TENGs can also operate as self-powered sensors, as variations in the amplitude, frequency, and waveform of their electrical output can directly reflect changes in external mechanical stimuli and surrounding environmental conditions. For instance, self-powered sliding and tactile sensors based on triboelectric signals have demonstrated high precision through frequency- and amplitude-dependent responses. Moreover, flexible and wearable TENGs have been shown to sensitively monitor physiological signals such as respiration, heart rate, pulse, and other biometric parameters by analysing their output waveforms [4]. In recent years, extensive efforts have been devoted to the design and optimization of TENGs to further enhance their performance. It is widely recognized that the electrical and mechanical characteristics of TENGs are primarily governed by device architecture, surface morphology, choice of triboelectric materials, and dielectric properties [5]. Based on structural configurations, TENGs are generally classified into four fundamental operating modes: contact–separation, lateral-sliding, single-electrode, and freestanding modes. Building upon these basic modes, more sophisticated and intelligent TENG architectures have been developed to meet diverse energy harvesting and self-powered sensing requirements across different application scenarios. The surface morphology of triboelectric materials plays a critical role in determining the effective contact area, which directly influences the amount of surface charge generated during contact electrification and, consequently, the overall output performance of TENGs. Equally important is the selection of appropriate fabrication materials, particularly triboelectric materials [6]. A larger difference in triboelectric polarity between the two contacting materials where one preferentially gains electrons and the other tends to lose electrons leads to enhanced electrical output. In addition, triboelectric materials with superior dielectric properties can suppress electrostatic breakdown while simultaneously increasing surface charge density. The intrinsic physical and chemical characteristics of the chosen materials further endow TENGs with desirable attributes such as flexibility, stretchability, and biocompatibility. Therefore, optimizing device structure, surface morphology, triboelectric material selection, and dielectric properties is essential for achieving significant improvements in both the mechanical robustness and electrical performance of triboelectric nanogenerators.

Nature provides a rich and powerful source of inspiration for both scientific exploration and technological innovation. Bioinspired design approaches, which emulate natural surfaces, structures, and materials [7], have proven to be highly effective in enhancing the performance of conventional devices and enabling the development of novel technologies. Representative examples include brittle star–inspired microlenses for advanced optical systems and gecko-inspired adhesive interfaces that allow wall-climbing by overcoming gravitational forces. In many cases, biomimetic applications can be realized using conventional fabrication tools and materials, particularly when the goal is to replicate specific biological functions such as sensing and signal processing. For example, multifunctional electronic skins capable of detecting temperature, strain, and touch are widely applied in robotics, while artificial neurons play an essential role in the advancement of deep learning and artificial intelligence systems. Among the numerous strategies aimed at improving the electrical and mechanical performance of triboelectric nanogenerators (TENGs), bioinspired designs have shown especially significant advantages. By mimicking natural structures, surface morphologies, material characteristics, and biological sensing or energy generation mechanisms, bioinspired TENGs exhibit substantial performance enhancements compared to conventional designs. Beyond improving output performance, these bioinspired approaches also provide greater adaptability across diverse operating environments and application scenarios an important aspect that is often overlooked. Moreover, the intrinsic advantages of TENGs, including their external stimulus responsiveness, high sensitivity, ease of fabrication and integration, tunable electrical output, and self-powered operation, enable their evolution into a wide range of applications, particularly in self-powered sensing systems. Biomimetic TENG-based applications, such as artificial electronic skin and neuromorphic devices, have driven notable progress in these fields and expanded opportunities for the development of low-cost artificial components that emulate human biological functions [3]. Accordingly, this chapter focuses on bioinspired TENGs and the biomimetic applications enabled by TENG technology, as illustrated in Fig. 2.1.

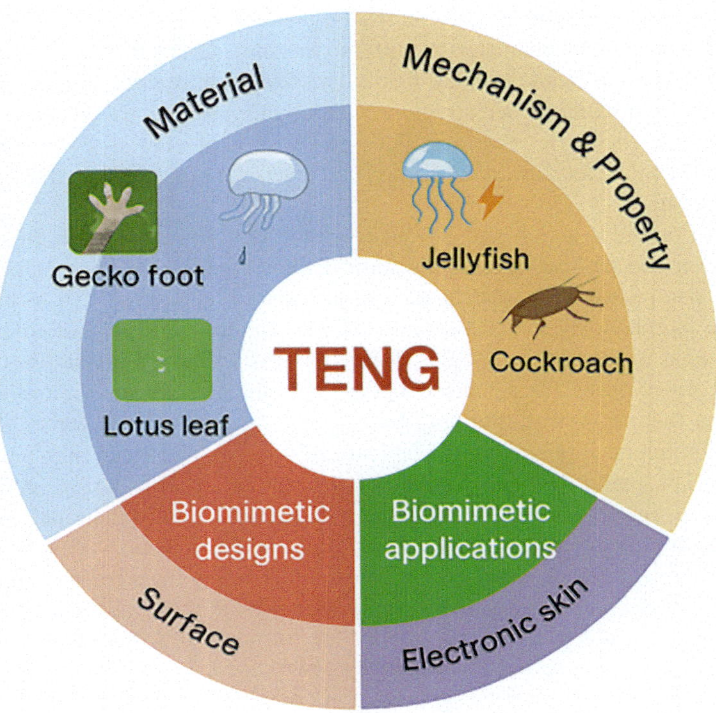

Fig. 2.1 Bioinspired designs and biomimetic applications of triboelectric nanogenerators

2.2 Bio-Inspired Designs of Triboelectric Nanogenerators

Structurally bionic TENGs represent an impressive combination of nanotechnology and bioinspired engineering in the rapidly developing field of energy harvesting. These TENGs, which are mostly separated into categories for plant and animal bionics, use the inherent qualities and structures of nature to increase their effectiveness and versatility. The complex patterns and designs found in a variety of flora, such as the vascular networks of plants or the microstructures of leaves, serve as inspiration for plant bionic TENGs. Improved energy conversion efficiencies result from the nanoscale replication of these natural templates to produce surfaces that optimize triboelectric phenomena. The textures, forms, and functions of animal tissues or organs are, nevertheless, replicated by animal bionic TENGs. These TENGs are made to efficiently capture and transform mechanical energy from motions or environmental interactions that are typical of the animal world, such as mimicking the skin, fur, or even the biomechanical movements of animals. In addition to expanding the use of TENGs in a variety of settings, this biomimetic approach creates new opportunities for the development of energy harvesters that are incredibly effective, environmentally friendly, and adaptable to a wide range of uses.

2.2.1 Plant-Inspired Structural Design

Zhang and co-workers [7] pioneered the development of a biomimetic triboelectric nanogenerator (bTENG) integrated with a novel motion-sensing system. This device effectively addresses a key limitation of conventional self-powered motion sensors used in wake-up circuits, which typically require relatively large mechanical forces to trigger activation. In contrast, the proposed bTENG is capable of responding to extremely subtle mechanical stimuli, enabling the detection of even minute disturbances. Drawing inspiration from plant morphology, the bTENG incorporates leaf-like tendril structures that significantly enhance its electrical output—achieving nearly a fourfold increase compared to conventional designs. This marked improvement greatly extends the detection range of wake-up circuits, thereby improving their sensitivity and reliability. Structurally, the bTENG consists of a base plane decorated with numerous flat, leaf-shaped tendrils. These tendrils are attached to an internal metal electrode and effectively increase the three-dimensional contact area during operation, leading to enhanced charge generation and higher electrical output. Each tendril is architecturally designed with a silicone rubber outer layer shaped like a leaf and embedded metal wires that mimic the vein structure found in natural leaves. This bioinspired composite configuration provides an optimal balance of mechanical properties, including high strength, appropriate stiffness, and excellent flexibility. As a result, the bTENG exhibits robust mechanical performance and long-term durability, making it well suited for practical applications that demand high sensitivity and stable operation. Yao et al. [8] have made significant progress in robotic tactile sensing by developing a self-powered electronic skin (e-skin) sensor based on a biomimetic triboelectric nanogenerator (TENG). In their design, the conical microstructure array found on the surface of *Calathea zebrine* leaves is replicated onto two polydimethylsiloxane (PDMS) friction layers, yielding a uniform and well-defined conical morphology. This bioinspired surface architecture is fabricated using a two-step molding process: first, a negative replica of the *C. zebrine* leaf surface is formed in PDMS, followed by a second molding step that transfers the conical microstructure array onto the PDMS substrate. The resulting microcones, with an average height of approximately 25.7 μm, a base diameter of about 25.4 μm, and an average spacing of roughly 33.6 μm, substantially increase the effective frictional contact area. Under applied pressure, these microstructures interlock between the opposing layers, thereby enhancing charge generation and improving the triboelectric output. This innovative replication strategy enables straightforward fabrication of a distinctive friction layer architecture for the TENG. Furthermore, the incorporation of polytetrafluoroethylene (PTFE) microfibres into the friction layers leads to a dramatic enhancement in pressure sensitivity, increasing it by up to 14-fold. Owing to its mechanical flexibility and structural adaptability, the self-powered TENG-based e-skin sensor can be readily integrated onto bionic or robotic hands, highlighting its strong potential for advanced robotic applications. By effectively combining biomimetic principles with nanotechnology, this work represents an important step forward in the development of high-performance, self-powered tactile sensors for next-generation robotic systems.

Wang et al. [9] successfully engineered a spatially multilevel nanofibrous membrane featuring a grid-like microstructure inspired by the natural architecture of frond leaves. The fabrication process involved the use of a metal mesh template combined with in situ polymerization and ultrasonic treatment to modify grid-structured polyurethane (PU) nanofibres with poly(3,4-ethylenedioxythiophene) (PEDOT) and carboxylated carbon nanotubes (CCNTs). The resulting biomimetic nanofibrous membrane exhibited exceptional sensing performance, including a high sensitivity of 5.13 kPa^{-1}, rapid response and recovery times of approximately 80 ms and 120 ms, respectively, and an ultralow pressure detection limit of 1 Pa. In addition to its sensing capability, the membrane demonstrated excellent versatility and scalability, enabling its use as a multifunctional platform for applications such as energy harvesting and electrothermal conversion. These results highlight the strong potential of this bioinspired approach for next-generation wearable devices and offer a novel pathway toward the development of high-sensitivity sensing systems. In a separate study, Wang and co-workers [10] developed a flexible plant-biomimetic hybrid self-powered sensor (PBHS) designed for the detection of motion spasms. The PBHS features a distinctive structural configuration, with surface patterns inspired by the leaf morphology of *Victoria amazonica*. These bioinspired surface textures, with crystalline feature sizes ranging from several hundred nanometers to a few micrometers, were fabricated using soft lithography transfer molding. The middle layer of the sensor incorporates layered poly(vinylidene fluoride–trifluoroethylene) (PVDF-TrFE) nanofiber (LPPN) sensors, which were produced via near-field electrospinning (NFES) by depositing approximately 2500 PVDF-TrFE piezoelectric polymer fibers onto a 108 mm × 12 mm flexible printed circuit board (FPCB). Encapsulation of these polymers with nano-microfibres (NMFs) significantly enhances the device's energy harvesting efficiency. This innovative PBHS integrates the LPPN sensors on the FPCB substrate with a biomimetic polydimethylsiloxane (PDMS)-based TENG, forming a hybrid energy harvesting and sensing system. The 2.2 mm thick LPPN wall, primarily composed of piezoelectric polymers, substantially improves the overall energy collection capability of the device. Notably, the output voltage of the hybrid sensor approaches 5 V, representing an enhancement of nearly 200% compared to the original PVDF-TrFE nanogenerator. Collectively, the structural design and functional performance of the PBHS mark a significant advancement in the development of efficient self-powered sensors, particularly for biomedical monitoring and related applications.

2.2.2 Animal-Inspired Structural Design

Wang et al. [11] creatively applied biomimetic principles in the development of a bionic-antennae-array (BAA) sensor inspired by the highly sensitive antennae of cockroaches. The core element of this biomimetic concept is an array of conductive fibres precisely arranged on the sensor surface to emulate the structure and sensing mechanism of cockroach antennae. Rather than serving merely as structural elements, these fibres are intentionally engineered to enhance the sensor's electrostatic sensing capability. A novel fabrication strategy is employed to coat the fibres with silver nanoparticles, with careful control over their distribution to maximize electrostatic balance and sensing sensitivity. This design choice closely mirrors the natural sensing mechanism of cockroach antennae, which detect environmental changes by responding to variations in surrounding electrostatic fields. A critical aspect of the BAA sensor lies in the fabrication of the polyamide silver (PA/Ag) fibres that form the antennae array. Precise regulation of fibre morphology and uniform dispersion of silver nanoparticles are essential for achieving optimal performance. By faithfully reproducing the operating principle of cockroach antennae namely, noncontact sensing through electrostatic field variations—the biomimetic fibres enable highly accurate detection of approaching objects. Consequently, the BAA sensor not only captures the structural essence of biological antennae but also replicates their functional behaviour, achieving high sensitivity and spatial resolution in motion detection through tailored fibre composition and arrangement. In a related effort, Zhu et al. [12] developed a self-powered bionic antenna (SBA) inspired by the structure and function of insect antennae to enhance tactile perception in robotic systems. This sensor is particularly suitable for micro-robotic applications, where lightweight design, high sensitivity, and real-time sensing both passive and active are critical requirements. The SBA exemplifies the effectiveness of biomimicry in engineering, incorporating a two-stage actuation mechanism that enables active sensing in both horizontal and vertical directions, closely mimicking the dynamic motion of insect antennae. Much like natural insect antennae that integrate multiple sensory functions, including the detection of mechanical stimuli, temperature, humidity, and chemical cues, the SBA is capable of actively interacting with its environment. Its flagellum-inspired sensing element responds to a variety of contact materials, analogous to the multiple receptors distributed along an insect antenna. Furthermore, the inclusion of muscle-like fibres, analogous to those driving antenna motion in insects, allows the SBA to achieve biomimetic movement and perception. By successfully reproducing both the structural and functional characteristics of biological antennae, this biomimetic SBA design underscores the strong potential of bioinspired strategies for advancing next-generation robotic tactile sensing systems.

Drawing inspiration from natural strategies that balance flexibility with protection, Niu et al. [5] reported the innovative design and fabrication of a bionic scales knitting triboelectric nanogenerator (BSK-TENG). By emulating the imbricated scale architectures found in animals such as snakes, fish, and pangolins, and integrating these bionic scales within a knitted textile framework, the BSK-TENG effectively addresses the challenge of incorporating complex and intelligent biomimetic structures into fabrics without compromising their inherent flexibility, comfort, or breathability. The device was realized using a novel high-speed V-bed flat knitting fully formed technique, which enables rapid, environmentally friendly production and makes the BSK-TENG suitable for large-scale manufacturing and commercial applications. The resulting wearable textile exhibits a distinctive three-dimensional hierarchical structure that is highly effective for harvesting biomechanical energy and water-drop energy, particularly in outdoor environments. Notably, the BSK-TENG also displays anisotropic bending behaviour, rendering it well suited for applications such as joint protection and supportive gear in outdoor sports. Inspired by the hierarchical structure of fish scales, Ma et al. [13] introduced a biomimetic fish-scale-like triboelectric nanogenerator (FSL-TENG). This device leverages the unique functional characteristics of fish scales to enable continuous energy harvesting from sliding and rotational motions in arbitrary directions, thereby avoiding inactive or "off" states. The FSL-TENG is fabricated using polyethylene terephthalate (PET) films and polytetrafluoroethylene (PTFE) sheets, which are patterned to replicate the overlapping geometry of natural fish scales. The design incorporates curvilinear features on both sides of the structure and a periodically symmetrical electrode arrangement, allowing efficient collection of mechanical energy from multidirectional motions. The electrode configuration is specifically inspired by the scale structure of the giant arapaima fish, known for its exceptional energy dispersion and protective capabilities. Experimental results demonstrate that the FSL-TENG achieves its highest current, voltage, and power output when oriented at a 45° angle, which corresponds to the maximum overlap area between the PTFE slider and the electrodes. Additionally, the PET film encapsulation enhances the device's flexibility and lightweight nature, broadening its applicability across a range of practical energy-harvesting scenarios.

Inspired by the unique microstructural features of shark skin, Yeh et al. [14] developed a wearable triboelectric sensor for gait analysis. This biomimetic approach employs a solid liquid TENG configuration in which a liquid metal is embedded within an Ecoflex elastomer surface patterned with a shark-skin-like microstructure. The key innovation lies in replicating the hydrophobic and ribbed microstructure of shark skin, which significantly enhances the sensor's sensitivity, durability, and long-term operational stability. Shark skin is naturally characterized by riblet-like surface features that reduce flow resistance and inhibit adhesion, and these properties are strategically translated into the sensor design to prevent unwanted liquid metal sticking during operation. Ecoflex, selected for its flexibility and biocompatibility, is micro structured to closely mimic shark skin, resulting in a hydrophobic, self-cleaning surface analogous to the lotus-leaf effect. Operating through a self-powered triboelectric mechanism, the interaction between the liquid metal and the micro structured Ecoflex surface enables efficient charge transfer, which is essential for accurate and reliable gait monitoring. Comprehensive experimental evaluations demonstrate that the sensor can effectively distinguish between different gait patterns and reliably differentiate healthy individuals from those with specific muscle impairments, highlighting its potential for wearable healthcare and rehabilitation monitoring. In a novel approach to human–machine interfaces (HMIs) for individuals with disabilities, Zhou et al. [15] introduced a bionic TENG-based sensor inspired by the croaking behaviour of frogs. This device integrates triboelectric sensing with a bioinspired mechanical amplification strategy to effectively capture muscle micromotions, particularly those of the masseter muscle. The sensor design emulates the coordinated motion of a frog's mouth and external vocal sac during croaking, where small muscle contractions induce large deformations of the vocal sac to amplify sound. Similarly, the proposed sensor consists of a flexible polydimethylsiloxane (PDMS) elastomer sensing layer coupled with a deformable vibrating film. This configuration converts subtle muscle movements into pronounced mechanical vibrations, thereby significantly amplifying the electrical signal. As a result, the sensor achieves a wide sensing range of 0–5 mm and a high output signal intensity of up to ±700 mV, corresponding to a signal enhancement approximately 206 times greater than that of conventional biopotential electromyography techniques. These features demonstrate the effectiveness of frog-inspired biomimetic designs in advancing high-performance, self-powered sensing technologies for next-generation HMI and biomedical applications.

2.2.3 *Additional Bioinspired TENGs*

This section highlights additional bioinspired triboelectric nanogenerators (TENGs) that do not fall neatly within the previously discussed categories. In these studies, inspiration is drawn not only from living biological systems but also from broader aspects of the natural environment and non-living phenomena. A key focus of this line of research is sustainability, with efforts directed toward recycling waste materials and reducing environmental burden by transforming discarded resources into functional energy-harvesting devices. Several pioneering studies have demonstrated that natural and recycled materials can be effectively repurposed as triboelectric components while maintaining the performance characteristics of conventional TENGs. For instance, recycled paper and waste polymers have been utilized to fabricate TENG-compatible materials that preserve efficient energy harvesting functionality. In another example, a solid–liquid contact TENG was developed based on the physical principle of a bubble moving through water, illustrating how non-biological natural mechanisms can also inspire innovative TENG designs. In one notable study, polymers derived from recycled paper were employed to fabricate a water-soluble and biodegradable TENG [16]. By combining cellulose nanocrystals and methylcellulose with polytetrafluoroethylene (PTFE), the researchers successfully converted paper waste into a functional, eco-friendly TENG. Remarkably, this water-soluble TENG exhibited a voltage output approximately three times higher than that of a conventional PTFE/Kapton-based TENG, demonstrating its strong potential as a sustainable solution to recycling challenges. Adopting a more targeted approach to addressing plastic waste, another study focused on fabricating TENGs from residual plastic polymers that are typically difficult or uneconomical to recycle [17]. In 2015 alone, approximately 150 megatons of plastic waste were generated from packaging, with less than 10% of plastics in the United States being recycled; the remainder were incinerated, landfilled, or released into the environment. In this work, commonly used packaging polymers including polypropylene (PP), polystyrene (PS), polyethylene (PE), and melamine formaldehyde (MF) were directly reused as triboelectric layers in environmentally friendly TENG devices. Because the recycled polymers required no additional chemical modification, the fabrication process was both simple and cost-effective. Among the tested materials, MF exhibited the strongest tribo-positive behaviour when paired with Kapton as the tribo-negative layer, leading to its selection for further evaluation in an optimized M-TENG configuration.

Under optimized operating conditions including load resistance, excitation frequency, and melamine formaldehyde (MF) layer thickness the M-TENG achieved a maximum output voltage exceeding 1500 V and a charge density of 1.176 $\mu C\ cm^{-2}$. The robustness of the device for both sensing and energy harvesting applications was further verified by its stable output voltage over more than 10,000 operating cycles. These results demonstrate that recycled MF is a highly effective triboelectric material and suggest that other recycled plastic polymers may exhibit comparable triboelectric performance, opening new avenues for sustainable TENG fabrication. While many bioinspired TENGs draw inspiration from living organisms such as plants and animals, some researchers have turned to non-living natural processes for innovative design concepts. One notable example is a bubble-based triboelectric nanogenerator (B-TENG) that harvests energy from the motion of an air bubble in water using a liquid–solid contact mechanism [18]. This frictionless energy-harvesting strategy was specifically developed to mitigate performance degradation caused by mechanical friction and repeated collisions between solid electrodes in conventional TENGs. As reported by Li et al. [18], liquid–solid contact TENGs (LS-TENGs) are also less susceptible to output fluctuations induced by changes in ambient air pressure and humidity. The proposed B-TENG effectively addresses the common challenge of incomplete separation between liquid and solid contact layers encountered in many LS-TENG designs, achieving a stable output voltage of approximately 17.5 V. Structurally, the two-electrode B-TENG consists of a 20 cm long polytetrafluoroethylene (PTFE) tube filled with distilled water, with two electrodes positioned at opposite ends one aluminium tape electrode ($20 \times 50\ mm^2$) and one 0.5 mm diameter wire electrode. Because both the PTFE tube and the water can carry triboelectric charges, the movement of the air bubble disrupts the electrostatic equilibrium between the two materials when the tube is tilted toward one electrode. This disturbance drives positive charge transfer from the external electrode on the tube to the internal electrode immersed in water. When the tube is tilted in the opposite direction and the internal electrode becomes enclosed by the bubble, the direction of charge transfer reverses. Performance evaluations showed that the B-TENG maintained stable output characteristics over at least 600 cycles, demonstrating its reliability for practical use. In addition to energy harvesting from water motion, the device also shows strong potential for water-level monitoring applications. Consequently, this unconventional, non-biological bioinspired TENG design offers promising opportunities for deployment in industrial environments, particularly in scenarios where hydrodynamic energy is typically wasted and continuous water-level sensing is required.

2.3 Biomimetic Applications

For decades, scientists have aimed to emulate human perceptual capabilities—such as environmental sensing, signal transmission, and neural information processing to advance intelligent robotics, artificial intelligence, and artificial organ technologies. In the biomedical field in particular, biomimetic devices that reproduce the functions of specific human tissues or organs have demonstrated remarkable potential for diagnostics, rehabilitation, and human–machine interaction. However, a critical limitation of many existing biomimetic systems is their reliance on rigid, bulky, and finite battery-based power supplies, which restrict long-term operation, flexibility, and practical deployment. To overcome these challenges and address the growing energy demand, there is a strong need for self-powered biomimetic systems that can harvest ambient energy while simultaneously enabling self-powered sensing. Among the various energy transduction technologies available, triboelectric nanogenerators (TENGs) stand out due to their unique advantages, including simple and scalable fabrication, low cost, high output voltage, exceptional sensitivity to external stimuli, mechanical flexibility and stretchability, ease of integration with diverse systems, and broad material compatibility. These attributes make TENGs particularly attractive for powering and sensing in next-generation biomimetic devices. A comprehensive comparison of recently developed material-based biomimetic TENGs is provided in Table 2.1.

Table 2.1 Detailed comparison of recent biomimetic material based triboelectric nanogenerators (TENGs)

Bionic objects	Devices	Working mode	Q_{SC}	I_{SC}	V_{OC}	Reference
Alathea zebrine leaf	TENG e-skin	CS	23.98 μC/m^2	6.29 nA	3.14	[8]
Nature frond leaf	TENG	CS	–	0.45 μA	78	[10]
Nymphaea tetragona	PBHS	CS	–	46 nA	13.1	[10]
Imbricate scales	BSK-TENG	SE	–	–	–	[5]
Shark skin	VCS-TENG	CS	–	0.7 μA	32	[14]
Lotus-leaf	bTENG	SE	–		1.2	[7]
Achilles tendon of Simmental cattle	PCOBE-TENG	SE	50 nC	5 μA	80	[19]
Bovine serum albumin	B-skin	SE	–	0.7 μA	474	[20]
Fibrous membrane	UVE-TENG	FT	–	1 μA	8 k	[21]

CS contact-separation mode, *SE* single electrode mode, *FT* freestanding triboelectric-layer mode (FT)

Q_{SC} short-circuit transferred charge, I_{SC} short-circuit current, V_{OC} open-circuit voltage

2.3.1 TENG Enabled Electronic Skin

As discussed earlier, the highly complex and hierarchical structures of human skin have provided valuable inspiration for the design of advanced triboelectric nanogenerators (TENGs) with superior performance and unique functionalities. At the same time, the intrinsic attributes of TENGs namely self-powered sensing, mechanical compliance, and seamless integrability—enable them to function effectively as artificial electronic skins capable of perceiving changes in the surrounding environment. To date, a wide variety of TENG-based electronic skin systems have been reported, and comprehensive reviews dedicated to this topic are already available in the literature [22]. Therefore, only a few representative and recent examples are highlighted here. Pu et al. [23] proposed an ultrastretchable, transparent, and soft triboelectric electronic skin designed for tactile sensing and biomechanical energy harvesting. By employing an ionic hydrogel as the electrode and an elastomer as the triboelectric layer, the device simultaneously achieved high optical transparency and exceptional stretchability. In addition to harvesting energy from human motion, the electronic skin exhibited a pressure sensitivity of 0.013 mV kPa^{-1}. As a proof of concept, a 3 × 3 pixel artificial skin array was demonstrated, capable of detecting both the magnitude and spatial location of applied pressure. Dong et al. [24] reported a stretchable and washable skin-inspired TENG (SI-TENG) that functions as an electronic skin. The device consists of a silicone rubber triboelectric layer combined with a silver-coated conductive yarn network electrode. Uniform infiltration of silicone rubber within the yarn network enabled good transparency while maintaining mechanical robustness. Owing to its high-pressure sensitivity, the SI-TENG-based electronic skin can serve as a wearable power source for harvesting biomechanical energy. Moreover, it can be integrated into intelligent prosthetic hands to recognize finger contact forces and detect weak physiological signals, such as vocal vibrations. To enable the detection of multidimensional tactile stimuli and gesture signals, Bu et al. [25] developed a stretchable triboelectric–photonic smart skin (STPS). This multifunctional device comprises a conductive silver nanowire (AgNW) network layer, a microcracked metal film, and a silicone rubber matrix incorporating aggregation-induced emission (AIE) luminophores. Upon stretching, the metal layer forms dispersed microcracks whose dimensions correlate with the applied strain, while the AIE-doped silicone rubber exhibits photoluminescence under light excitation. The triboelectric sensing component, formed by the silicone rubber and AgNW electrode, demonstrates an exceptionally high-pressure sensitivity of 34 mV Pa^{-1} in the low-pressure regime and is particularly responsive to vertical pressure stimuli.

2.3.2 TENG Enabled Artificial Neuromorphic Devices

A typical neuromorphic process involves the perception of external stimuli by mechanoreceptors, followed by signal transduction and information processing through synapse-like systems. The development of self-powered artificial mechanoreceptors, autonomous signal transduction without external power supplies, and integrated artificial neuromorphic devices is therefore highly beneficial for applications in artificial intelligence, human–machine interfaces, and assistive technologies for individuals with disabilities. For patients suffering from irreversible hearing loss caused by the degeneration of hair cells, cochlear implants remain one of the most effective clinical solutions. In this context, artificial basilar membranes (ABMs) have attracted increasing interest for hearing-assistive devices because they can emulate key cochlear functions, including tonotopic frequency mapping, acoustic-to-electrical energy conversion performed by hair cells, and the passive frequency selectivity of the natural basilar membrane [26]. To replicate cochlear tonotopy, Jang et al. [27] proposed a triboelectric-based artificial basilar membrane (TEABM). The TEABM consists of an array of beams with different lengths, where each beam is tuned to respond to a specific resonant frequency. When exposed to acoustic excitation, contact electrification between a Kapton film and an aluminium foil enables direct conversion of sound waves into electrical signals. In this design, the apical region of the membrane responds preferentially to high-frequency sounds, while the basal region is more sensitive to low-frequency stimuli, closely mimicking the frequency-dependent response of the human cochlea. The TEABM demonstrates a frequency selectivity spanning from 294.8 to 2311 Hz, and its sensitivity to incident sound pressure increases with the number of channels, reaching values in the range of 1.74 to 13.1 mV Pa^{-1}.

The development of intelligent tactile sensors capable of directly interacting with neural systems represents a fundamental step toward realizing artificial nervous systems. Inspired by the mechanisms of somatosensory signal generation and neuroplasticity-driven signal processing, Wu et al. [28] reported a biomimetic neuromorphic tactile sensor with intrinsic learning and memory functions based on a carefully engineered single-electrode triboelectric nanogenerator (TENG). In this design, a polyimide–reduced graphene oxide (PI:rGO) hybrid layer serves as the negative triboelectric material. The embedded rGO sheets act as effective charge-trapping sites, enabling spontaneous charge migration within the hybrid layer and thereby increasing the overall triboelectric charge density. A defining feature of this neuromorphic tactile sensor is that its electrical output increases progressively with repeated pressing, analogous to the excitatory postsynaptic potential observed in biological synapses. From a bionic standpoint, this gradually enhanced output reflects a learning-like behaviour, as it encodes information about both the current stimulus and the history of prior mechanical inputs. Moreover, the temporal decay of the output signal due to charge dissipation closely resembles the human "forgetting curve," characterized by an initial gradual decline followed by a more rapid drop, further reinforcing the biological relevance of the device's signal dynamics. In a related study, Liu et al. [29] developed a self-powered artificial synapse transistor (SPST) by integrating a TENG with an electric double-layer organic field-effect transistor. In this system, the TENG functions as a tactile perception unit that generates pre-synaptic voltage spikes in response to external mechanical stimuli, while the source–drain channel of the transistor serves as the post-synaptic element. The TENG output is coupled to the gate electrode, effectively acting as pre-synaptic stimulation. Upon mechanical activation, the SPST exhibits output behaviours that closely mimic key biological synaptic functions, including excitatory postsynaptic current (EPC), paired-pulse facilitation (PPF), dynamic filtering, and the transition from short-term to long-term synaptic plasticity. The pronounced hysteresis observed in the device arises from the high proton conductivity of the chitosan-based gate electrolyte as the gate voltage is swept forward and backward. Beyond accurately reproducing fundamental synaptic behaviours such as EPC and PPF, this biomimetic system also emulates complex sensory functions observed in nature, such as a web-weaving spider's ability to perceive airflow variations and vibrations generated by trapped prey.

2.4 Conclusion and Future Perspectives

Bioinspired triboelectric nanogenerator (TENGs) hold substantial promise for advancing biotechnology and related interdisciplinary fields. Unique features observed in humans, animals, and plants have already been shown to translate effectively into applications in artificial intelligence, biomonitoring, and energy harvesting. When exploring new strategies to modify and enhance TENG performance, drawing inspiration from nature whether through biological structures, functional mechanisms, or naturally derived materials has proven to be a powerful and versatile approach. As a result, bioinspired TENGs introduce a range of advantageous characteristics, including improved sweat resistance, enhanced adaptability, expanded output tunability, and broader application potential compared to conventional designs. Importantly, bioinspired approaches also open pathways toward environmentally sustainable alternatives to traditional TENG materials, many of which are non-biodegradable and potentially harmful. Natural materials such as wood, recycled paper, and even plastic waste have already been explored as triboelectric components, representing meaningful progress toward addressing plastic pollution and improving the environmental footprint of TENG technologies. Nevertheless, despite these advances, significant challenges remain. Many bioinspired TENGs that incorporate natural materials still rely on non-biodegradable or plastic components, meaning that the issue of electronic waste is not yet fully resolved. To truly mitigate this problem, future bioinspired TENGs should aim to be fabricated entirely from natural and biodegradable materials. Achieving this goal will require persistent efforts to identify environmentally benign materials for every component of the device, including both positive and negative triboelectric layers, electrodes, and structural elements. If natural substitutes can be found for all components, fully biodegradable TENG systems rather than partially sustainable ones will become feasible. In parallel, advancing research on the effective recycling of electronic devices can further support the development of bioinspired TENGs. At present, recycling remains challenging because device materials and components are often tightly integrated and difficult to separate. Addressing this limitation will require the development of new recycling strategies or innovative TENG fabrication methods that enable easier disassembly and material recovery at the end of a device's life cycle. Continued exploration of the natural world is also essential for uncovering new inspiration for next-generation TENG applications. This may involve studying a wider diversity of flora and fauna or conducting deeper investigations into selected species to identify previously unrecognized features that can be translated into device design. Undiscovered micro- and nanostructures could further enhance surface charge density, while unique macrostructures may inspire entirely new TENG architectures, such as systems modelled after hummingbird wings or fish gills. As highlighted in this review, organisms that thrive in environments vastly different from human conditions such as marine species or flying animals are particularly valuable sources of inspiration. Given that much of the ocean remains unexplored, continued research into marine organisms may yield transformative insights for bioinspired TENG development. Moreover, many biological systems can withstand extreme temperatures and harsh environments, offering valuable design principles for TENGs intended to operate under extreme conditions. Fungi represent another promising yet underexplored source of inspiration. They have already been used to degrade pollutants, decompose biomaterials, and modify polymeric substances, suggesting their potential role in the development of TENG-compatible materials that can be safely decomposed after use. Owing to their unique biochemical and structural properties, fungi may also inspire novel TENG designs beyond sustainability considerations alone. Ideally, future studies reporting new TENG technologies should include a dedicated discussion on recyclability or biodegradability, supported by environmental testing that evaluates natural degradation pathways or environmentally safe disposal methods. Furthermore, substantially more research is needed on human-inspired TENGs, particularly for applications in robotics and artificial intelligence. Among the various bioinspired domains reviewed, human-inspired TENGs remain the least explored. Applying both simple and complex aspects of human physiology such as joint lubrication mechanisms or the alveolar structure of the lungs could inspire innovative nanogenerator designs with advanced functionality. Overall, the pioneering developments summarized in this work represent only the beginning of a rapidly evolving and highly promising field. Bioinspired TENGs are poised to play a transformative role in future nanotechnologies, bridging sustainability, intelligence, and advanced energy harvesting in unprecedented ways.

References

1. Wu C, Wang AC, Ding W, Guo H, Wang ZL (2019) Triboelectric nanogenerator: a foundation of the energy for the new era. Adv Energy Mater 9:1802906
2. Xia R, Zhang R, Jie Y, Zhao W, Cao X, Wang Z (2022) Natural cotton-based triboelectric nanogenerator as a self-powered system for efficient use of water and wind energy. Nano Energy 92:106685
3. Zhang Z, Qi Z, Sun X, Xu J (2023) A triboelectric nanogenerator based on bionic design for harvesting energy from low-frequency vibration. Int J Non Linear Mech 157:104540
4. Zheng Q, Shi B, Li Z, Wang ZL (2017) Recent progress on piezoelectric and triboelectric energy harvesters in biomedical systems. Adv Sci 4(7):1700029. https://doi.org/10.1002/advs.201700029
5. Niu L, Peng X, Chen L, Liu Q, Wang T, Dong K, Pan H, Cong H, Liu G, Jiang G et al (2022) Industrial production of bionic scales knitting fabric-based triboelectric nanogenerator for outdoor rescue and human protection. Nano Energy 97:107168
6. Zou Y, Raveendran V, Chen J (2020) Wearable triboelectric nanogenerators for biomechanical energy harvesting. Nano Energy 77:105303. https://doi.org/10.1016/j.nanoen.2020.1053
7. Zhang C, Dai K, Liu D, Yi F, Wang X, Zhu L, You Z (2020) Ultralow quiescent power-consumption wake-up technology based on the bionic triboelectric Nanogenerator. Adv Sci 7:2000254
8. Yao G, Xu L, Cheng X, Li Y, Huang X, Guo W, Liu S, Wang ZL, Wu H (2020) Bioinspired triboelectric Nanogenerators as self-powered electronic skin for robotic tactile sensing. Adv Funct Mater 30:1907312
9. Wang M, Dong L, Wu J, Shi J, Gao Q, Zhu C, Morikawa H (2022) Leaf-meridian bio-inspired nanofibrous electronics with uniform distributed microgrid and 3D multi-level structure for wearable applications. NPJ Flex Electron 6:34
10. Wang J, Chen CC, Shie CY, Li TT, Fuh YK (2022) A hybrid sensor for motor tics recognition based on piezoelectric and triboelectric design and fabrication. Sens Actuators A Phys 342:113622
11. Wang F, Ren Z, Nie J, Tian J, Ding Y, Chen X (2020) Self-powered sensor based on bionic antennae arrays and triboelectric nanogenerator for identifying noncontact motions. Adv Mater Technol 5:1900789
12. Zhu D, Lu J, Zheng M, Wang D, Wang J, Liu Y, Wang X, Zhang M (2023) Self-powered bionic antenna based on triboelectric nanogenerator for micro-robotic tactile sensing. Nano Energy 114:108644
13. Ma G, Li B, Niu S, Zhang J, Wang D, Wang Z, Zhou L, Liu Q, Liu L, Wang J et al (2022) A bioinspired triboelectric nanogenerator for all state energy harvester and self-powered rotating monitor. Nano Energy 91:106637
14. Yeh C, Kao F-C, Wei P-H, Pal A, Kaswan K, Huang Y-T, Parashar P, Yeh H-Y, Wang T-W, Tiwari N et al (2022) Bioinspired shark skin-based liquid metal triboelectric nanogenerator for self-powered gait analysis and long-term rehabilitation monitoring. Nano Energy 104:107852
15. Zhou H, Li D, He X, Hui X, Guo H, Hu C, Mu X, Wang ZL (2021) Bionic ultra-sensitive self-powered electromechanical sensor for muscle-triggered communication application. Adv Sci 8:2101020
16. Wang T, Li S, Tao X, Yan Q, Wang X, Chen Y, Huang F, Li H, Chen X, Bian Z (2022) Fully biodegradable water-soluble triboelectric nanogenerator for human physiological monitoring. Nano Energy 93:106787
17. Geyer R, Jambeck JR, Law KL (2017) Production, use, and fate of all plastics ever made. Sci Adv 3(7):e1700782. https://doi.org/10.1126/sciadv.1700782
18. Li CZ, Liu XY, Yang DF, Liu Z (2022) Triboelectric nanogenerator based on a moving bubble in liquid for mechanical energy harvesting and water level monitoring. Nano Energy 95:106998. https://doi.org/10.1016/j.nanoen.2022.106998

19. Song B, Fan X, Shen J, Gu H (2023) Ultra-stable and self-healing coordinated collagen-based multifunctional double-network organohydrogel e-skin for multimodal sensing monitoring of strain-resistance, bioelectrode, and self-powered triboelectric nanogenerator. Chem Eng J 474:145780
20. Leng Z, Zhu P, Wang X, Wang Y, Li P, Huang W, Li B, Jin R, Han N, Wu J et al (2023) Sebum-membrane-inspired ProteinBased bioprotonic hydrogel for artificial skin and human-machine merging Interface. Adv Funct Mater 33:2211056
21. Yang P, Shi Y, Tao X, Liu Z, Li S, Chen X, Wang ZL (2023) Self-powered virtual olfactory generation system based on bionic fibrous membrane and electrostatic field accelerated evaporation. EcoMat 5:e12298
22. Chen H, Song Y, Cheng X, Zhang H (2019) Self-powered electronic skin based on the triboelectric generator. Nano Energy 56:252–268
23. Pu X, Liu M, Chen X, Sun J, Du C, Zhang Y, Zhai J, Hu W, Wang ZL (2017) Ultra stretchable, transparent triboelectric nanogenerator as electronic skin for biomechanical energy harvesting and tactile sensing. Sci Adv 3:e1700015
24. Dong K, Wu Z, Deng J, Wang AC, Zou H, Chen C, Hu D, Gu B, Sun B, Wang ZL (2018) A stretchable yarn embedded triboelectric nanogenerator as electronic skin for biomechanical energy harvesting and multifunctional pressure sensing. Adv Mater 30:1804944
25. Bu T, Xiao T, Yang Z, Liu G, Fu X, Nie J, Guo T, Pang Y, Zhao J, Xi F, Zhang C, Wang ZL (2018) Stretchable triboelectric-photonic smart skin for tactile and gesture sensing. Adv Mater 30:1800066
26. Song WJ, Jang J, Kim S, Choi H (2014) Piezoelectric performance of continuous beam and narrow supported beam arrays for artificial basilar membranes. Electron Mater Lett 10:1011–1018
27. Jang J, Lee J, Jang JH, Choi H (2016) A triboelectric-based artificial basilar membrane to mimic cochlear tonotopy. Adv Healthc Mater 5:2481–2487
28. Wu C, Kim TW, Park JH, Koo B, Sung S, Shao J, Zhang C, Wang ZL (2020) Self powered tactile sensor with learning and memory. ACS Nano 14:1390–1398
29. Liu Y, Zhong J, Li E, Yang H, Wang X, Lai D, Chen H, Guo T (2019) Self-powered artificial synapses actuated by triboelectric nanogenerator. Nano Energy 60:377–384

Chapter 3
Multifunctional Bio-Inspired Energy Harvesters: Harnessing Piezoelectric, Wind, and Acoustic Energy

Abstract Multifunctional energy harvesters that can transform a variety of ambient energies into useable electricity have been developed as a result of the search for sustainable, decentralized power sources. The design and integration of multifunctional energy harvesting systems that include piezoelectric, wind, and acoustic energy conversion processes are examined in this chapter using bio-inspired approaches. These hybrid gadgets, which are inspired by natural systems, show improved efficiency and flexibility. Innovative materials are discussed, such as bio-derived piezoelectric polymers, biomimetic surfaces for controlling wind flow, and acoustic metamaterials modelled after natural auditory organs. Alongside developments in nanostructured interfaces and self-powered sensor networks, system designs for synergistic energy coupling and storage are examined. These harvesters provide prospective avenues for self-sustaining microelectronic systems, smart infrastructures, and autonomous environmental monitoring platforms by combining biomimicry with multi-modal energy conversion.

Keywords Piezoelectric energy · Acoustic energy · Wind energy · Biomimicry

3.1 Introduction

Due to the finite supply of conventional energy and the rapid development of total energy consumption, energy disasters are now among the most important issues [1]. Sustainable, clean, and pollution-free green energy sources capable of delivering output powers in the range of microwatts to milliwatts are becoming increasingly vital in today's technology-driven world. Such compact yet efficient power generators are essential for reliably and rapidly charging the growing class of portable, wearable, and smart electronic devices. Although batteries and capacitors have become significant energy sources in recent years, their short lifespan and challenging disposal pose serious problems for long-term uses and environmental safety. In today's rapidly evolving electronics landscape, piezoelectric, pyroelectric, and

M. Y. Mir, J. A. Parray, *Nanoenergy Production*, SpringerBriefs in Energy,
https://doi.org/10.1007/978-3-032-20859-0_3

triboelectric nanogenerators have emerged as some of the most promising strategies for green and sustainable energy harvesting. In particular, extensive research in recent years on piezoelectric nanogenerators (PNGs) and triboelectric nanogenerators (TENGs) has demonstrated their ability to convert biomechanical and ambient mechanical stimuli into usable electrical energy. A key direction for future high-tech electronics lies in the development of flexible, smart PNG devices capable of harvesting energy through multiple modes while employing environmentally friendly and non-toxic materials. The integration of biocompatible piezoelectric materials into flexible and wearable multifunctional devices represents a significant step toward reducing electronic waste an escalating environmental concern that affects both ecosystems and human health. Such materials enable the creation of safe, lightweight, and degradable energy harvesters, offering a sustainable alternative to traditional battery-powered electronics. Given their inherent advantages and versatility, biocompatible piezoelectric materials capable of multiple energy-collection modes present a highly feasible pathway toward next-generation green electronics. As emphasized by Yuan et al. [2], mechanical energy is one of the most abundant and widely accessible ambient energy sources, available through countless mechanical and biomechanical activities. Harnessing this ubiquitous resource holds tremendous potential for producing clean and renewable power for future self-powered systems. In our daily lives, this ongoing energy harvesting source manifests itself in a variety of ways, including walking, running, joint movements, kicking, driving, and taking. Additionally, the frequency of sound waves, sometimes referred to as acoustic energy, is crucial for producing electricity [3]. Additionally, a variety of waste noises from autos, human voices, electronic gazettes, instruments, and enterprises are significant sources of alternative energy that might be transformed into green energy for the benefit of humanity. Furthermore, the spontaneous availability of wind and water waves in nature makes the design of an energy collecting PNG device to produce electricity from them very commendable. Therefore, the successful integration of multiple green energy sources such as mechanical vibrations, sound waves, water-wave motion, and wind flow into a single multifunctional device opens up new possibilities for powering a wide range of smart technologies without imposing any ecological burden. A lightweight and highly sensitive energy-harvesting platform of this kind would be particularly valuable for advancing robotics, artificial sensing systems, and both human and animal healthcare monitoring. Owing to these advantages, piezoelectric nanogenerators (PNGs) are gaining increasing attention as a promising solution to meet the growing energy demands of wearable, implantable, wireless, and portable self-powered devices. Although piezoelectric polymers generally exhibit lower piezoelectric coefficients compared to widely used inorganic materials such as PZT, $BaTiO_3$, $ZnSnO_3$, $NaNbO_3$, ZnO, GaN, and MoS_2, they remain highly attractive for nanogenerator (NG) design. Their unique benefits including non-toxicity, intrinsic flexibility, biocompatibility, chemical robustness, mechanical durability, and ease of processing make them especially suitable for the next generation of flexible and environmentally friendly energy-harvesting systems. These attributes position piezoelectric polymers as key materials for multifunctional, sustainable, and user-friendly electronics of the future. To overcome the limitations associated with inorganic piezoelectric materials particularly their brittleness and toxicity polymer based piezoelectric nanogenerators (PNGs) have emerged as a promising platform for the efficient design of next-generation energy-harvesting devices. Despite significant progress, only a limited number of studies have reported devices capable of integrating multiple energy collecting modes while still maintaining high output performance. This gap presents a valuable opportunity to develop biocompatible, flexible, and multifunctional polymer-based PNGs that offer superior peak power density and enhanced energy-conversion efficiency for applications in smart green electronics and e-healthcare monitoring. Among various polymer candidates, poly (vinylidene fluoride) (PVDF) was chosen for this study because of its excellent piezoelectric coefficient, strong mechanical durability, ferroelectric characteristics, and desirable flexibility [4]. These properties make PVDF particularly suitable for constructing hybrid energy harvesters capable of responding to multiple ambient stimuli. Table 3.1 provides an overview of bio-inspired hybrid energy harvesting systems that combine piezoelectric, wind, and acoustic modalities.

3.2 Bio-Inspired Piezoelectric Energy Harvester

A piezoelectric vibration energy harvester operates on the fundamental piezoelectric effect exhibited by piezoelectric materials [13]. Figure 3.1 illustrates the diverse multifunctional roles of piezoelectric energy-harvesting systems, emphasizing their applications in biosensing, self-powered implantable devices, motion-driven wearable electronics, and electrostimulation-assisted regenerative technologies. When subjected to external environmental vibrations, the piezoelectric element—the core energy conversion component—undergoes mechanical deformation along with the vibrating structure. This deformation induces a relative displacement of the internal positive and negative charge centres within the material, leading to charge accumulation on its top and bottom surfaces [14]. The resulting charge separation generates a potential difference between the electrodes, driving electron flow through an external circuit. In this manner, vibrational mechanical energy is effectively converted into usable electrical energy. By adding nonlinear effects, piezoelectric energy harvesting technology greatly enhances the vibration amplitude, extending the strain and stress range of piezoelectric materials and producing greater output levels. This technology is especially well-suited for low-frequency vibration situations because to its high output voltage, quick reaction time, and simplicity of miniaturization [15]. In the world of energy harvesting, piezoelectric vibration energy harvesters are distinctive due to these features. Because of their high degree of environmental flexibility and effective energy conversion mechanism, bio-inspired structures have drawn a lot of interest. The effectiveness and stability of energy harvesting may be further increased by combining bio-inspired structures with piezoelectric energy harvesting [16]. The piezoelectric bio-inspired nonlinear energy harvester shows great application potential by fusing the special benefits of

Table 3.1 Representative multifunctional bio-inspired energy harvesters integrating piezoelectric, wind and acoustic mechanisms

S. no	System	Primary mechanism(s)	Biomimetic inspiration	Output	Reference
1	Bionic-dipteran bistable piezo harvester	Piezoelectric (bistable)	Mimics dipteran wing skeleton to enable low-frequency bistability	Ultralow-frequency operation (e.g., ≈4 Hz) with measurable mW/μW power under lab excitation	[5]
2	Bionic cochlear triboelectric acoustic sensor (film-TENG)	Triboelectric (acoustic frequency selectivity)	Basilar-membrane/cochlea inspired electrode distribution for frequency-selective acoustic sensing	Frequency-selective response 20–3000 Hz demonstrated	[6]
3	Bi-stable piezoelectric broadband harvester (Venus-flytrap motif)	Piezoelectric (bi-stable, broadband)	Inspired by rapid shape change of Venus-flytrap to obtain broadband response	Broadband response and improved harvested power vs. comparable linear designs	[7]
4	Piezoelectric–triboelectric hybrid (synergistic design)	Hybrid PENG + TENG	Synergistic material/electrode design to combine both effects for stronger output	Reported improved combined output vs. single-mode devices	[8]
5	Hybrid bistable swing-impact PT (tribo + piezo) for low-freq	Hybrid triboelectric + piezoelectric	Mechanical frequency-up/impact conversion inspired by mechanical oscillators in nature	Tunable potential barrier, frequency-up conversion, broadband low-frequency harvesting	[9]
6	Plant-protein/biodegradable triboelectric TENGs (Agri applications)	Triboelectric (biodegradable materials)	Uses plant-protein films and mulch-film concepts (bio-based TENG)	Demonstrated biodegradable TENG materials and practical farm/greenhouse demo prospects	[10]
7	Silk fibroin and nacre-like silk membranes	Osmotic energy/piezoelectric	Nacre-like, silk-crosslinked membranes for osmotic and mechanical energy harvesting	Osmotic energy harvesting and robust membrane performance	[11]
8	Collagen/peptide assemblies (piezoelectric biopolymers)	Piezoelectric	Molecular engineering of peptide/collagen assemblies to induce strong polarization	Demonstrated coin-sized generators and enhanced piezoelectric response	[12]

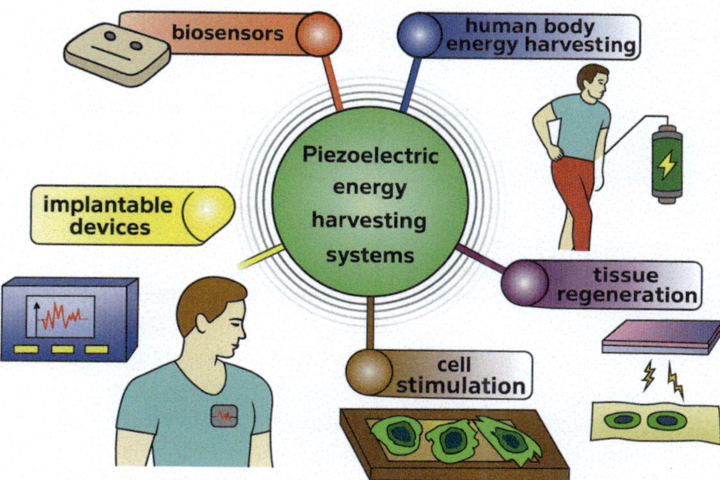

Fig. 3.1 Representative applications of piezoelectric materials in energy harvesting and biomedical engineering

bio-inspired structures with the superior performance of piezoelectric materials. Researchers have now created a variety of piezoelectric bio-inspired nonlinear energy harvesters, each with unique material choices, geometric shapes, and operational mechanisms. They offer a dependable way to power microelectronic devices by effectively converting vibration energy into electrical energy and adapting to a variety of complicated vibration conditions [17].

A bistable energy-harvesting structure can be created by hinging together two flexible beams with two rigid links, forming a mechanism that closely resembles the skeletal configuration of dipteran wings. When the attached verification mass is excited by substrate vibrations, the piezoelectric patch bonded to the flexible beam undergoes deformation and generates electrical energy. Dynamic analyses of displacement responses and output voltages under various conditions show that inter-well vibrations can be activated under both low damping and strong excitation. Importantly, the highest harvesting efficiency is achieved when the system operates in high-amplitude inter-well vibration modes. Experimental results demonstrate that a maximum output power of 0.143 mW is obtained in this mode at an excitation frequency of 4 Hz and an acceleration of 0.7 g ($g = 9.8\ m/s^2$). The system also exhibits the ability to harvest energy under ultra-low-frequency and mild excitation conditions. Building on the benefits of nonlinear dynamics, Li and Jing [18] developed and analysed an innovative coupled vibration energy-harvesting system designed to exploit the combined effects of structural coupling and nonlinear behaviour. By integrating a nonlinear X-shaped structure with a piezoelectric harvester through two distinct mounting configurations horizontal and vertical they significantly improved the operational bandwidth relative to traditional cantilever-type harvesters, including those supported by spring–mass systems. The X-shaped structure enables the harvester to function efficiently over a much broader frequency range, extending its performance into the ultra-low-frequency regime, where conventional designs typically underperform.

Inspired by biological systems, several innovative bio-inspired piezoelectric vibration energy harvesters have been developed to effectively convert low-frequency and random vibrations into electrical energy. Drawing on the concept of plant parasitism, Fu et al. [19] proposed a host–parasite heuristic piezoelectric harvester that synergistically integrates biostability with frequency-up-conversion mechanisms to harvest irregular low-frequency vibrations. In their design, a compression beam is driven into a buckled state by clamping both ends and applying axial compressive stress. The device exploits a "plucking" interaction in which low-frequency vibrations of the primary (host) beam excite high-frequency resonant vibrations in a parasitic beam, significantly improving energy-conversion efficiency. Both electromechanical modelling and experimental validation showed that the harvester can accumulate more than 1 mJ of energy within 100 s under pseudo-random vibration inputs, demonstrating excellent broadband low-frequency harvesting capability. Inspired by the natural opening mechanism of soybean pods, Hao et al. [20] introduced a novel bionic bistable piezoelectric energy harvester constructed from metallic nanomaterials. The device features a gradient nanostructure produced via surface mechanical attrition treatment, which enhances its mechanical robustness and energy-harvesting performance. Experimental and numerical analyses confirmed substantial improvements in voltage output and operational bandwidth under both intermittent and continuous inter-well vibration modes. Potential applications include smart monitoring systems embedded in tires, vehicle bridge interfaces, and suspension components. Wang et al. [21] developed another biologically inspired design a dual-band piezoelectric energy harvester modelled after the unique structure of a woodpecker's head, particularly the dual-branch configuration of its tongue bone. Their device incorporates a piezoelectric array optimized through finite element analysis to determine the most effective placement of piezoelectric patches within the array. To accommodate the multiple alternating current outputs, an improved rectification circuit was engineered. The harvester supports two operational modes, motion mode and impact mode allowing it to adapt to a wide range of human activities. This makes the system particularly suitable for self-powered smartwatches, wristbands, and other wearable electronics.

3.3 Wind Energy Harvesting

In the field of wind-driven PZT energy harvesting, plant-inspired designs have been widely explored to enhance efficiency by mimicking leaf geometry and natural fluttering dynamics. Wang et al. [22] developed leaf-shaped PZT harvesters that utilize vortex-induced vibrations (VIVs), drawing inspiration from the venation patterns of dicot leaves. By incorporating venation geometries such as isosceles trapezoid structures, the design improved the transmission of aerodynamic forces to the piezoelectric material. Wind tunnel experiments demonstrated a peak output power of 0.85 μW at a wind speed of 10 m s^{-1}, with open-circuit RMS voltages ranging from 0.54–1.26 V at 6 m s^{-1} and 0.76–1.69 V at 11 m s^{-1}. Li et al. [23] further advanced leaf-mimetic concepts by designing stalk–blade harvesters capable of operating under both cross-flow and parallel-flow fluttering. Their device used a flexible beam functioning as an artificial petiole to emulate natural leaf flutter. Peak power outputs of 0.02 mW and 0.21 mW were recorded for parallel- and cross-flow conditions, corresponding to power densities of 0.09 and 0.871 mW cm^{-3}, respectively. A biomorph stalk variant exhibited even stronger performance, reaching 0.61 mW and 1.3 mW cm^{-3} due to increased vortex-induced stresses. Abdelkareem [24] introduced a hybrid harvesting platform that integrated artificial leaves with PZT sheets to simultaneously harness both wind and solar energy. The design was highly scalable—ten-leaf configurations generated up to 3.42 mW by increasing the number of stacked PZT layers. The same research group also developed a bio-inspired rectenna shaped like a leaf for radio-frequency energy harvesting. Operating in the 1.6–2.65 GHz range, the rectenna achieved an output of 960 mV and 0.42 mW, demonstrating the versatility of plant-inspired structures for multifunctional ambient energy harvesting.

Lastly, Wang et al. [25] developed a bio-inspired artificial palm leaf harvester incorporating an embedded PZT transducer, modeled on the bending–twisting flutter of natural palm leaves. Under airflow, the structure reproduced coupled deformation modes and generated 0.76 μW at 3 m s^{-1}, while wind-tunnel tests demonstrated a stable flutter frequency of 3.56 Hz and a low cut-in speed of ~2 m s^{-1}. Hobeck and Inman [26] designed a "deploy-and-forget" PZT harvester optimized for turbulent, low-velocity conditions by mimicking the passive flow-driven oscillations of grass blades. Their system consists of vertically oriented, cantilevered beams arranged in a multi-element array, each integrating a surface-bonded PZT layer. Random vortex and turbulence induced vibrations excite each cantilever independently, and the resulting surface strain on the PZT layers enables continuous kinetic-to-electrical energy conversion. Large deflection under low aerodynamic forces is made possible by the flexible substrate's improved compliance. The system produced up to 1 mW of output per beam during wind tunnel tests at 11.5 m s^{-1}, confirming its viability for scalable, maintenance-free EH under erratic flow circumstances. Researchers have also developed a range of bluff-body geometries inspired by trees, flowers, butterflies, and honeycombs to enhance wind-induced PZT power generation. Hobbs and Hu [27] introduced tree-trunk-shaped, multi-arrayed bluff bodies equipped with PZT harvesters, drawing inspiration from the sway of tree trunks under wind loading. Mounted on a platform that permitted only cross-flow tilting, the system showed that the foremost cylinder produced the lowest energy output suggesting it could be replaced with a non-functional dummy to reduce fabrication costs. The authors further proposed exploiting cooperative interaction mechanisms, analogous to energy-saving behaviours observed in animal groupings; for example, fish swimming in schools can harvest vortex energy generated by upstream individuals. A large cluster of trunk-like harvesters could therefore leverage similar aerodynamic interactions to significantly improve PZT energy harvesting performance. For more efficient PZT wind harvesting, Wang et al. [28] developed a butterfly-inspired bluff body designed to exploit both vortex-induced vibrations (VIVs) and the galloping effect across different wind-speed regimes. The vertically oriented wing-like profile promotes early flow separation and rapid vortex shedding at low wind speeds, initiating VIV, whereas flow asymmetry triggers galloping oscillations at higher speeds. The coexistence of these two excitation mechanisms expands the harvester's effective operational bandwidth [29]. The same design principles are relevant to underwater hydroelectric energy harvesting, where similar fluid–structure interactions govern device performance. Once the wind speed exceeds the critical threshold, the butterfly-shaped bluff body undergoes self-excited oscillation, dynamically deforming the supporting cantilever and generating strain in the attached PZT layer. Compared with conventional cuboid bluff bodies, the butterfly geometry induces greater aerodynamic lift and drag, leading to higher energy conversion efficiency. The vertical butterfly configuration demonstrated the best performance, achieving an RMS voltage of 5.5 V at 3 m s^{-1} surpassing both the horizontal butterfly design (1.94 V) and the cuboid bluff body (3 V) [30].

3.4 Acoustic Energy Harvesting

Advanced acoustic energy harvesters (AEHs) can be broadly classified into three major types based on their working mechanisms: electromagnetic generators (EMGs), triboelectric nanogenerators (TENGs), and piezoelectric generators [31]. EMGs, although well established in traditional energy harvesting, perform poorly in acoustic environments because the low acoustic energy density and rapid fluctuations in sound pressure make it difficult for the moving conductor to effectively cut through magnetic field lines [32]. TENGs, on the other hand, leverage the synergistic action of electrostatic induction and the triboelectric effect to deliver remarkably high electrical outputs. Despite this advantage, their performance is highly sensitive to environmental factors such as temperature and humidity, which influence electrostatic charge retention. Moreover, TENG devices typically rely on repeated contact separation cycles between two materials with different triboelectric polarities. This requirement demands precise control of the separation distance, increasing fabrication complexity, cost, and the likelihood of mechanical failure when exposed to high vibrations or sudden shocks. In contrast, piezoelectric generators have become the most prevalent AEH technology due to several key benefits: they exhibit high sensitivity even to subtle mechanical disturbances, feature a simple capacitor-like structure, are easy and inexpensive to manufacture, and maintain excellent stability under diverse environmental conditions. These characteristics make piezoelectricity a highly reliable and practical foundation for next-generation AEH systems. The inherent piezoelectric effect of piezoelectric materials is often used by piezoelectric generators to produce charges by the mechanical deformation of orientated molecular dipoles. Because sound has a low acoustic energy density, piezoelectric-based AEHs must operate in resonant mode in order to provide the maximum amount of electrical output. However, the limited frequency spectrum of the piezoelectric generator is its main drawback. When an acoustic wave's frequency deviates from the AEH's prescribed resonance frequency, the generator's performance deteriorates dramatically.

Most studies on piezoelectric generator based acoustic energy harvesters (AEHs) have focused on boosting electrical output, primarily by amplifying the incoming sound pressure—one of the most effective strategies for enhancing energy conversion. To harvest low-frequency acoustic energy, researchers employed multiple PVDF beam arrays integrated into a quarter-wavelength acoustic resonator. Their system produced a response voltage of 1.48 V, corresponding to a volume power density of 0.002 μW cm^{-3}, under an incident sound pressure level (SPL) of 110 dB at 146 Hz. This approach enables even rigid piezoelectric materials such as lead zirconate titanate (PZT) ceramics to efficiently vibrate at low frequencies, thereby improving AEH performance. Generally, three types of acoustic structures are used to amplify sound pressure: Helmholtz resonators, half-wave/quarter-wave tube resonators, and acoustic metamaterials. For instance, Yang et al. [33] designed a compliant-top-plate Helmholtz resonator coupled with dual PZT cantilever beams to achieve strong multimode interactions between the resonator and the piezoelectric elements. Their AEH system delivered an output power of 0.137–1.430 mW (0.035–0.36 μW cm^{-3}) at 170–206 Hz and 100 dB SPL. Similarly, researchers integrated multiple PZT cantilever plates into a quarter-wavelength tube resonator to develop a highly efficient low-frequency AEH. The device achieved a maximum total output voltage of 5.089 V, equivalent to 1.367 μW cm^{-3}, when operated at 199 Hz under a 100 dB SPL. In another approach, Yuan et al. [34] utilized a PZT patch bonded to a metallic substrate equipped with a proof mass to form a locally resonant acoustic metamaterial capable of converting incident sound waves into electrical energy. Their system generated an electrical output of 210 μW (0.011 μW cm^{-3}) at 155 Hz and 114 dB SPL. According to the published study, even if a Helmholtz or tube resonator could significantly increase the SPL of the incident sound in the resonator, there were still drawbacks due to the device's enormous volume, high weight, intricate structure, and manufacturing process. Additionally, high-power piezoelectric AEHs continue to operate at a particular resonance frequency and very high SPL, making them better suited for usage in high SPL, single, and specific-frequency sound sources. Therefore, collecting multi-frequency acoustic energy at the lowest feasible SPL is an essential development direction for the piezoelectric generator-based AEHs in order to achieve a high-efficiency acoustoelectric conversion in the real noisy environment. Piezoelectric polymers offer several advantages over traditional rigid ceramic piezoelectrics, including low weight, high flexibility, low acoustic impedance, and ease of processing into various shapes and sizes. Their ability to undergo significant deformation under low-frequency acoustic waves makes them highly suitable for practical, large-scale deployment in real-world environments [35]. However, their direct use in high-power AEH applications remains limited due to their inherently low electromechanical coupling efficiency. Recent progress in polymer nano structuring has shown that converting piezoelectric polymers into nanofibers can significantly enhance their electrical output. For example, Mahanty et al. [36] used PVDF–Mg nanofibers as the active layer in an AEH and achieved a maximum power density of 8.36 μW (0.28 μW cm^{-2}) at a resonant frequency of 126 Hz under 120 dB SPL. Similarly, the development of two-dimensional MOF-modified PVDF nanofibers enabled the fabrication of a flexible, highly sensitive nanogenerator, which exhibited an improved power density of 6.25 μW (1.04 μW cm^{-2}) and an acoustic sensitivity of 0.95 V Pa^{-1} at 110 dB SPL and 120 Hz. Using polyacrylonitrile (PAN) nanofibers, remarkably high output power of 210.3 μW (17.53 μW cm^{-2}) at 230 Hz and 117 dB SPL is reported [37], demonstrating the significant potential of polymer nanofiber systems. Notably, most AEHs based on piezoelectric organic polymer nanofibers operate as non-resonant systems. Their enhanced acoustoelectric performance primarily arises from the ultrathin nanofiber structures, which can easily vibrate or deform under acoustic excitation, leading to efficient energy conversion even without resonance-assisted amplification. Additionally, it was observed that such a large power was measured at a very low frequency and typically dropped off rapidly as the frequency increased. The majority of piezoelectric polymer nanofiber-based AEHs still lack the power of piezoelectric AEHs that use resonators and PZT ceramics, but they do have the advantages of a simpler, lighter, and lower-volume device structure and manufacturing process. The presented work on piezoelectric polymer-based AEHs focused less on the device construction and associated vibration mode and more on enhancing the polymer's piezoelectric performance.

3.5 Conclusion and Future Perspectives

The practical viability of translating a piezoelectric nanogenerator (PNG) from laboratory-scale demonstrations to real world applications can be evaluated by considering several essential device parameters. These include design strategies and scalability, sensing performance, simplicity and speed of energy harvesting, long-term mechanical stability, durability, and the availability of diverse mechanical energy sources. When these factors are optimized, as in the case of the proposed multi-source PNG (MSPNG), the device can achieve high sensitivity, strong energy-conversion efficiency, and robust output performance. Consequently, the MSPNG shows strong potential as a cost-effective and versatile alternative energy solution capable of harvesting energy from multiple ambient sources. Drawing from the extensive experimental results previously discussed, we propose a realistic outdoor application scenario for the MSPNG along a highway. In such an environment, several forms of mechanical energy such as pressure from passing vehicles, natural and vehicle induced wind, environmental sound, rainfall impact, and pedestrian footsteps are continuously wasted. Integrating the MSPNG into the highway infrastructure would allow simultaneous harvesting of these mechanical stimuli. The harvested and stored energy could then be used to power a wide range of low-power electronic systems, including mobile device charging modules, LED lighting, traffic signal indicators, environmental monitoring sensors, and even security or structural-health-monitoring devices. This multi-functional capability highlights the MSPNG's promise for sustainable, self-powered infrastructure in smart transportation and urban environments.

References

1. Madruga S (2021) Modeling of enhanced micro-energy harvesting of thermal ambient fluctuations with metallic foams embedded in phase change materials. Renew Energy 168:424–437
2. Yuan M, Cao Z, Luo J, Chou X (2019) Recent developments of acoustic energy harvesting: a review. Micromachines 10(1):48
3. Panicker SS, Rajeev SP, Thomas V (2023) Impact of PVDF and its copolymer-based nanocomposites for flexible and wearable energy harvesters. Nano Struct Nano Objects 34:100949
4. Hwang YJ, Choi S, Kim HS (2019) Highly flexible all-nonwoven piezoelectric generators based on electrospun poly (vinylidene fluoride). Sens Actuators A Phys 300:111672
5. Zhou J, Zhao X, Wang K, Chang Y, Xu D, Wen G (2021) Bio-inspired bistable piezoelectric vibration energy harvester: design and experimental investigation. Energy 228:120595
6. Liu Y, Zhu Y, Liu J, Zhang Y, Liu J, Zhai J (2018) Design of bionic cochlear basilar membrane acoustic sensor for frequency selectivity based on film triboelectric nanogenerator. Nanoscale Res Lett 13(1):191
7. Khatua DK, Kim SJ (2022) A high performance piezoelectric–triboelectric hybrid energy harvester by synergistic design. Energy Adv 1(9):613–622
8. Li J, Chen J, Guo H (2021) Triboelectric nanogenerators for harvesting wind energy: recent advances and future perspectives. Energies 14(21):6949. https://doi.org/10.3390/en14216949
9. Chen W, He Z, Zhao J, Mo J, Ouyang H (2024) Hybrid triboelectric-piezoelectric energy harvesting via a bistable swing-impact structure with a tuneable potential barrier and frequency-up conversion effects. Appl Energy 375:124123
10. Jiang C, Zhang Q, He C, Zhang C, Feng X, Li X, Zhao Q, Ying Y, Ping J (2022) Plant-protein-enabled biodegradable triboelectric nanogenerator for sustainable agriculture. Fundam Res 2(6):974–984
11. Xin W, Xiao H, Kong XY, Chen J, Yang L, Niu B, Qian Y, Teng Y, Jiang L, Wen L (2020) Biomimetic nacre-like silk-cross linked membranes for osmotic energy harvesting. ACS Nano 14(8):9701–9710. https://doi.org/10.1021/acsnano.0c01309
12. Bera S, Guerin S, Yuan H et al (2021) Molecular engineering of piezoelectricity in collagen-mimicking peptide assemblies. Nat Commun 12:2634. https://doi.org/10.1038/s41467-021-22895-6
13. Hu Y, Xue HA, Hu HP (2007) A piezoelectric power harvester with adjustable frequency through axial preloads. Smart Mater Struct 16:1961–1966. https://doi.org/10.1088/0964-1726/16/5/054
14. Askari M, Ghandchi Tehrani M, Brusa E, Carrera A, Delprete C (2025) Design, fabrication and evaluation of a quad-finger multimodal vibration energy harvester utilizing MFC generators. Mech Adv Mater Struct:1–21. https://doi.org/10.1080/15376494.2025.2465911
15. Li H, Tian C, Deng ZD (2014) Energy harvesting from low frequency applications using piezoelectric materials. Appl Phys Rev 1:041301. https://doi.org/10.1063/1.4900845
16. Cha Y, Hong J, Lee J, Park J-M, Kim K (2016) Flexible piezoelectric energy harvesting from mouse click motions. Sensors (Basel) 16:1045
17. Zhang Y, Liu C, Jia B, Ma D, Tian X, Cui Y et al (2024) Kirigami-inspired, three-dimensional piezoelectric pressure sensors assembled by compressive buckling. NPJ Flexible Electron 8:23. https://doi.org/10.1038/s41528-024-00310-6
18. Li M, Jing XJ (2019) Novel tunable broadband piezoelectric harvesters for ultralow-frequency bridge vibration energy harvesting. Appl Energy 255:113829. https://doi.org/10.1016/j.apenergy.2019.113829

19. Fu H, Sharif-Khodaei Z, Aliabadi F (2019) A bio-inspired host-parasite structure for broadband vibration energy harvesting from low-frequency random sources. Appl Phys Lett 114:143901. https://doi.org/10.1063/1.5092593
20. Hao F, Wang B, Wang X, Tang T, Li Y, Yang Z et al (2022) Soybean-inspired nanomaterial-based broadband piezoelectric energy harvester with local biostability. Nano Energy 103:107823. https://doi.org/10.1016/j.nanoen.2022.107823
21. Wang B, Long Z, Hong Y, Pan Q, Lin W, Yang Z (2021) Woodpecker-mimic two-layer band energy harvester with a piezoelectric array for powering wrist-worn wearables. Nano Energy 89:106385. https://doi.org/10.1016/j.nanoen.2021.106385
22. Wang W, Wang X, He X, Wang M, Shu H, Xue K (2019) Comparisons of bioinspired piezoelectric wind energy harvesters with different layout of stiffeners based on leaf venation prototypes. Sens Actuators A Phys 298:111570
23. Li S, Yuan J, Lipson H (2011) Ambient wind energy harvesting using cross-flow fluttering. J Appl Phys 109(2)
24. Abdelkareem MAA, Jing X, Ahmed Ali MK, Choy Y (2025) Recent advances in vibration energy harvesting reinforced by bioinspired designs/structures. Smart Mater Struct 34(8):083002. https://doi.org/10.1088/1361-665X/adfb37
25. Wang K, Xia W, Ren J, Yu W, Feng H, Hu S (2023) Wind energy harvesting inspired by palm leaf flutter: observation, mechanism and experiment. Energ Conver Manage 284(2):116971. https://doi.org/10.1016/j.enconman.2023.116971
26. Hobeck JD, Inman DJ (2012) Artificial piezoelectric grass for energy harvesting from turbulence-induced vibration. Smart Mater Struct 21:105024. https://doi.org/10.1088/0964-1726/21/10/105024
27. Hobbs WB, Hu DL (2012) Tree-inspired piezoelectric energy harvesting. J Fluids Struct 28:103–114. https://doi.org/10.1016/j.jfluidstructs.2011.08.005
28. Wang J, Zhang C, Gu S, Yang K, Li H, Lai Y, Yurchenko D (2020) Enhancement of low-speed piezoelectric wind energy harvesting by bluff body shapes: spindle-like and butterfly-like cross-sections. Aerosp Sci Technol 103:105898. https://doi.org/10.1016/j.ast.2020.105898
29. Tang L, Zhao L, Yang Y, Lefeuvre E (2015) Equivalent circuit representation and analysis of galloping-based wind energy harvesting. IEEE ASME Trans Mechatron 20:834–844
30. Yang X, He X, Li J, Jiang S (2019) Modeling and verification of piezoelectric wind energy harvesters enhanced by interaction between vortex-induced vibration and galloping. Smart Mater Struct 28(11):13. https://doi.org/10.1088/1361-665X/ab4216
31. Liu Q, Li Q, Fang Z, Zhou X, Wang R, Zuo C (2021) Piezoelectric energy harvesting for flapping wing micro air vehicle and flapping wing sensing based on flexible polyvinylidene fluoride. Appl Sci 11(3):1166. https://doi.org/10.3390/app11031166
32. Chen H, Xing C, Li Y, Wang J, Xu Y (2020) Triboelectric Nanogenerators for a macro-scale blue energy harvesting and self-powered marine environmental monitoring system. Sustain Energy Fuels 4:1063–1077
33. Yang A, Li P, Wen Y, Lu C, Peng X, He W, Zhang J, Wang D, Yang F (2014) Note: high-efficiency broadband acoustic energy harvesting using Helmholtz resonator and dual piezoelectric cantilever beams. Rev Sci Instrum 85(6):066103. https://doi.org/10.1063/1.4882316
34. Yuan M, Cao Z, Luo J, Zhang J, Chang C (2017) An efficient low-frequency acoustic energy harvester. Sens Actuators A Phys 264:84–89
35. Ramadan KS, Sameoto D, Evoy S (2014) A review of piezoelectric polymers as functional materials for electromechanical transducers. Smart Mater Struct 23(3):033001. https://doi.org/10.1088/0964-1726/23/3/033001
36. Mahanty B, Ghosh SK, Jana S, Roy K, Sarkar S, Mandal D (2021) All-fiber acousto-electric energy harvester from magnesium salt-modulated PVDF nanofiber. Sustain Energy Fuels 5(4):1003–1013
37. Roy K, Jana S, Mallick Z, Ghosh SK, Dutta B, Sarkar S, Sinha C, Mandal D (2021) Two-dimensional MOF modulated fiber nanogenerator for effective acoustoelectric conversion and human motion detection. Langmuir 37(23):7107–7117. https://doi.org/10.1021/acs.langmuir.1c00700

Chapter 4
Bio-Inspired Gel Materials: Unlocking New Potentials in Advanced Energy Technologies

Abstract Bio-inspired gel materials have emerged as a versatile class of soft, functional systems that mimic the structural, mechanical, and physicochemical attributes of natural tissues. Their unique combination of high-water content, tunable conductivity, biocompatibility, and adaptive responsiveness enable advanced functions that traditional synthetic materials often cannot achieve. Recent advancements in molecular design, cross-linking strategies, and the incorporation of conductive polymers, nanomaterials, and biomolecules have positioned these gels as key enablers for next-generation energy technologies. By bridging principles from biology, materials science, and energy engineering, bio-inspired gels offer a transformative pathway toward sustainable, flexible, and high-performance energy systems. The chapter concludes with perspectives on emerging trends, challenges, and opportunities for integrating these materials into scalable, environmentally responsible energy platforms.

Keywords Conductive polymer gels · Hydrogel · Organogel · Supercapacitors · Ion thermoelectric system

4.1 Introduction

Cross linked network materials that can exhibit properties spanning from soft and compliant to stiff and mechanically robust are collectively known as gels. Conventionally, most gels particularly hydrogels are constructed from homopolymer or copolymer chains interconnected through either chemical bonds or physical interactions [1]. In physical gels, the network is formed via non-covalent interactions, where polymer chains associate over finite regions rather than through discrete point-to-point bonds, leading to the formation of junction zones. Typical interactions responsible for such physical crosslinking include hydrogen bonding, hydrophobic associations, and electrostatic (charge-based) interactions. In contrast, gels formed through covalent crosslinking are commonly referred to as chemical gels. These systems generally possess superior mechanical strength and long-term

M. Y. Mir, J. A. Parray, *Nanoenergy Production*, SpringerBriefs in Energy,
https://doi.org/10.1007/978-3-032-20859-0_4

stability owing to the robustness of covalent bonds. Chemical crosslinking can be achieved through various strategies, such as high-energy irradiation, reactions between complementary functional groups, radical polymerization processes, or the incorporation of multifunctional crosslinking agents. Recent advances in gel chemistry and fabrication techniques have expanded the range of building blocks beyond traditional polymer chains. A wide spectrum of materials including metals, large organic molecules, carbon-based nanostructures, and other inorganic components can now be assembled into three-dimensional gel networks using approaches such as self-assembly, template-assisted methods, and tailored crosslinkers. These building units may span different dimensionalities, from zero-dimensional quantum dots [2] to one-dimensional nanowires [3] and two-dimensional nanosheets [4]. The resulting gel networks are formed through chemical or physical interactions either among the building blocks themselves or between the building blocks and added crosslinkers. Common driving forces in these systems include covalent bonding, hydrogen bonding, hydrophobic interactions, Coulombic forces, and π–π interactions. Gels especially hydrogels have been widely investigated and applied in bio-related fields such as tissue engineering, controlled drug delivery, biomedical sensing, and diagnostic imaging [5]. Their success in these applications arises from the unique combination of chemical composition and physical architecture inherent to gel networks [6]. While gels strongly absorb and retain solvents, their crosslinked framework prevents dissolution, maintaining structural integrity. Their mechanical properties can be finely tuned over a broad range, and the porous network allows solvent and small molecules to diffuse through the polymer chains. Moreover, the high surface area and intrinsic biocompatibility of many gel matrices enable the incorporation and uniform distribution of biologically active entities such as cells, enzymes, and therapeutic agents via physical entrapment or chemical immobilization, yielding multifunctional hybrid materials. The shape, internal structure, and molecular architecture of gels can also be precisely controlled, allowing tailored mechanical, chemical, and transport properties. Importantly, through rational molecular design, gel networks can be engineered to interact intelligently with their environment. As a result, a wide variety of stimuli responsive gels have been developed that can sense and respond to external triggers such as temperature, light, pH, and mechanical stress.

4.2 Conductive Polymer Gels

Conductive polymers including polypyrrole (PPy), polyaniline (PANI), and poly(3,4-ethylenedioxythiophene) (PEDOT) belong to a class of conjugated π-polymers characterized by alternating single and double covalent bonds along their molecular backbone (Table 4.1) [7]. This conjugated structure enables the delocalization of π-electrons over extended chain segments, which is the fundamental origin of their electrical conductivity. The unsaturated backbone facilitates the movement of these delocalized electrons, thereby forming efficient pathways for charge transport and enabling the flow of mobile charge carriers through the material. Incorporating conductive polymers into hydrogel-based sensors offers several

Table 4.1 Comparison of conductive gels based on appearance, conductivity, advantages, and disadvantages

Conductive materials	Appearance	Conductivity (S/cm)	Advantages	Disadvantages	References
Polyaniline	Black, Green	$1.0 \times 10^{-5} - 0.43$	High electrical conductivity, intrinsic antibacterial activity, excellent mechanical properties promote cell adhesion, growth and proliferation	Strict and carefully controlled synthesis conditions	[10]
Polypyrrole	Black	0.195 – 0.8	High electrical conductivity, good biocompatibility, low cost, high chemical and structural stability	Poor mechanical strength and low solubility	[11]
PEDOT: PSS	Bluish-black	$1.0 \times 10^{-2} - 40$	High electrical conductivity, good biocompatibility, promote cell adhesion and proliferation, high chemical and structural stability	Susceptible to deformation	[12]
Graphene oxide	Black	$2.0 \times 10^{-2} - 6.62 \times 102$	High electrical conductivity, good biocompatibility, excellent mechanical properties, low cost	Complex synthetic procedures, GO is difficult to uniformly disperse during hydrogel synthesis	[13]
Carbon nanotubes	Black	1.7 – 20	High electrical conductivity, excellent mechanical properties low cost	Complex synthetic methods, which increase fabrication difficulty and reduce scalability. Carbon nanotubes (CNTs) are not easily dispersed during hydrogel synthesis, tending to agglomerate	[14]
MXene	Black	$1.0 - 1.0 \times 10^{3}$	High electrical conductivity, biocompatible, excellent mechanical properties, self-healing capability	Complex synthetic procedures. MXene is difficult to uniformly disperse during hydrogel synthesis	[15]

(continued)

Table 4.1 (continued)

Conductive materials	Appearance	Conductivity (S/cm)	Advantages	Disadvantages	References
AgNPs	Grayish white,	$1.0 \times 10^{-4} - 0.58$	High electrical conductivity, antibacterial activity, good biocompatibility, simple synthesis process, high chemical and structural stability	Risk of reactive oxygen species generation	[16]
AuNPs	Faint yellow, brown	$8.0 \times 10^{-4} - 1.0 \times 10^{-2}$	High electrical conductivity antibacterial properties biocompatible, high chemical and structural stability	Complex synthetic procedures with a potential risk of reactive oxygen species generation	[17]
Mussel-inspired (catechol / PDA-based) conductive hydrogels	Dark brown–black	0.1 – 10	Strong wet adhesion, self-healing, good stretchability, bio-inspired catechol chemistry	Catechol oxidation reduces long-term performance; nanofiller aggregation; biocompatibility concerns at high filler loadings	[18]
Chitosan-based conductive hydrogels	Light yellow to dark gray/ black	$\sim 10^{-3} - 1$	Biodegradable, biocompatible, tunable mechanical softness, inexpensive	High filler loading decreases porosity; lower conductivity under wet conditions; potential cytotoxicity depending on filler	[19]

distinct advantages. Firstly, their presence significantly reduces electrical resistance by promoting efficient charge transport, including the movement of solvent ions within the hydrogel network. This enhanced ionic electronic conduction improves signal transduction and sensor sensitivity. Secondly, conductive polymer chains exhibit excellent mechanical flexibility and strong compatibility with conventional polymer matrices, allowing them to be seamlessly integrated into hydrogel systems. Unlike rigid metallic fillers or stiff inorganic nanoparticles, conductive polymers can reinforce the hydrogel without compromising its softness and deformability, thereby improving tensile strength, durability, and overall mechanical performance while maintaining flexibility. As a result, one of the most promising materials for creating high-performance hydrogel sensors is conductive polymers. Conductive polymers are commonly used as the polymer framework or as conductive fillers in conductive hydrogels in electrocardiograms. When conductive polymers are simply introduced as fillers into a hydrogel matrix, they often aggregate into discontinuous, island-like domains within the porous network. This non-uniform dispersion can interrupt charge-transport pathways and, consequently, limit the overall electrical conductivity of the hydrogel. As a result, simultaneously achieving high electrical performance and strong mechanical integrity remains a significant challenge. One of the most effective strategies to enhance the conductivity of polymer hydrogels is to directly dope or chemically cross-link conductive polymers into the hydrogel network, thereby minimizing or eliminating the presence of electrically insulating components. For example, Zhou et al. developed a conducting polymer hydrogel (CPH) by doping and cross-linking polypyrrole (PPy) during in situ polymerization using tannic acid (TA) and Fe^{3+} ions [8]. In this system, PPy serves not only as the conductive phase but also as the primary structural framework of the hydrogel. Fe^{3+} ions act as oxidizing agents to initiate pyrrole polymerization and simultaneously function as ionic crosslinkers by coordinating with TA, leading to the formation of an ionically cross-linked network. The hydroxyl groups present on the benzene rings of TA interact with PPy chains through intermolecular electrostatic interactions, creating additional crosslinking points, while protonation of the nitrogen atoms along the PPy backbone imparts electrical conductivity. Consequently, TA plays a dual role as both a dopant and a crosslinker, resulting in electrically conductive gels with significantly enhanced electronic conductivity. Moreover, the mechanical strength and electrical properties of these conductive polymer hydrogels can be precisely tailored by adjusting the concentration of the dopant/cross-linking agents. By varying the TA content, CPHs with three distinct formulations were obtained, exhibiting conductivities in the range of 0.05–0.18 S cm^{-1}. In situ polymerization also helps mitigate polymer chain entanglement, a common issue in conductive polymer composites, by promoting uniform growth and integration of conjugated polymer networks. Using this approach, Huang and co-workers fabricated a hydrogel electrolyte by polymerizing polyaniline (PANI) in situ within a dual-network hydrogel composed of cross-linked poly (vinyl alcohol) (PVA) and polyacrylamide/acrylic acid (PAM/AA) [9]. The resulting PANI–PPG-based supercapacitors demonstrated characteristic capacitive behaviour, which was attributed to the high pseudo capacitance and conductivity of the in situ polymerized PANI electrodes, coupled with the abundant three-dimensional ionic transport channels provided by the polymer hydrogel electrolyte.

4.3 Energy Applications of Conductive Polymer Gels

Supercapacitors frequently use conductive polymers, a significant class of pseudocapacitive materials that capitalize on quick and reversible electron exchange reactions at or close to the electrode surface. With mechanical characteristics akin to those of normal polymers, these conjugated polymers exhibit high gravimetric and volumetric pseudo capacitance in a variety of nonaqueous electrolytes, which may make them valuable for the creation of lightweight and flexible devices. However, during high-rate cycling, bulk conductive polymers have short cycle lifetimes and fast decaying capacitances. This could be due to the volume changes that occur during the charge/discharge operations and the conductivity drop brought on by the doping state changes. Conductive polymer gels (CPGs) with three-dimensional (3D) interconnected network architectures are widely employed to overcome these challenges because they facilitate efficient electron transport, provide short diffusion pathways for electrolyte ions to access electroactive sites, and possess a porous framework that can accommodate volume changes during charge discharge processes. In addition, the doping states in CPGs are relatively stable owing to strong chemical interactions between the dopant molecules and the polymer chains. As a result, dopant molecule crosslinked CPGs have demonstrated outstanding electrochemical performance in supercapacitor applications. For instance, Pan et al. [20] reported that a polyaniline (PANI) hydrogel delivered a high specific capacitance of approximately 480 F g^{-1} at a current density of 0.2 A g^{-1} in a 1 M H_2SO_4 electrolyte. Notably, the capacitance decreased by only about 7% when the current density was increased tenfold, highlighting the excellent rate capability of the PANI hydrogel. This superior performance was attributed to its highly conductive network and well-developed porous structure. Similarly, a polypyrrole (PPy) hydrogel crosslinked with copper phthalocyanine tetrasulfonate (CuPcTs) exhibited a markedly enhanced specific capacitance compared with pristine PPy without crosslinkers [21]. The CuPcTs-doped PPy hydrogel achieved a specific capacitance of nearly 400 F g^{-1} at 0.2 A g^{-1}, whereas pristine PPy showed only 232 F g^{-1}. This improvement arose from increased electrical conductivity, a one-dimensional morphology, and a more open porous architecture. Beyond electrochemical performance, CPGs are particularly attractive for flexible supercapacitors because they inherit favourable mechanical properties from polymeric materials, including high elasticity and flexibility. When synthesized using strategies such as interfacial polymerization, the tailored microstructures of CPGs can impart enhanced mechanical strength and resilience, further improving device performance under deformation. Shi et al. [22] demonstrated this concept by fabricating PPy hydrogels with structure-induced elasticity and employing them as electrodes in all-solid-state supercapacitors. The resulting devices exhibited a high specific capacitance of 380 F g^{-1} along with excellent mechanical adaptability. Even under severe bending, the capacitance showed minimal variation, which was attributed to the porous network accommodating deformation of the PPy backbone and the strong adhesion between the PPy hydrogel and the current collector.

Lithium ion batteries (LIBs) dominate the consumer electronics market owing to their high energy storage capacity, high energy efficiency, and relatively low weight, making them more compact and portable than supercapacitors. Despite these advantages, LIBs suffer from inherent limitations, particularly restricted cycle life and comparatively slow charge–discharge rates. Since the 1980s, conductive polymers have been explored as both anode and cathode materials in LIBs. Their appeal arises from the reversible interactions between lithium ions and the conjugated polymer backbones, which enable lithium insertion and extraction during electrochemical cycling. This reversible redox behaviour allows conductive polymers to actively participate in energy storage processes, offering an alternative to conventional inorganic electrode materials and motivating continued research into polymer based electrodes for advanced LIB systems. The electrochemistry and material design of conductive polymers for electrode applications have been thoroughly reviewed [23]. However, in reduced forms, conductive polymers typically exhibit low conductivity and poor durability, which significantly limits their potential uses. Conductive polymers are being used as electrode materials in LIBs and are also showing promise as binder materials. Every active particle in a battery needs to be precisely sized, formed, and connected to the solid or liquid electrolyte and the current collector in order to optimize energy utilization [24]. Nevertheless, conventional binder systems used in LIB electrodes are typically binary hybrid systems, in which the conductive additive is randomly dispersed within an insulating polymer matrix. This random distribution often leads to electrical bottlenecks, poor interparticle contacts, and limited accessibility of certain electrode components. As a result, electrical conductivity and mechanical integrity can deteriorate during prolonged cycling. In contrast, conductive polymers offer a dual-function advantage by integrating the roles of both binder and conductive additive. By synergistically combining the mechanical robustness of traditional polymers with the electronic conductivity of organic conductors, conductive polymers can suppress aggregation of conductive additives and prevent delamination of the polymer layer during repeated charge–discharge cycles. More importantly, the chemical structures of conductive polymers are highly tunable, allowing their physical and chemical properties to be tailored through rational modification of polymer chains. For example, Wu et al. [25] adopted a systematic strategy involving chemical synthesis, quantum calculations, and comprehensive spectroscopic and mechanical characterization to design a polymer binder with simultaneously high electronic conductivity and strong mechanical integrity. An especially promising approach for enhancing LIB performance is the use of conductive polymer gels (CPGs) as binder materials. The continuous and highly conductive three-dimensional CPG network provides efficient electron transport pathways, direct electrical contact with the current collector, and robust interparticle connectivity, leading to excellent rate capability. In addition, the hierarchical porous structure of CPG-based electrodes maximizes the interfacial area between active materials and electrolytes while facilitating rapid ion diffusion. Owing to their hydrophilic nature, hydrogels can strongly adhere to the surfaces of active particles, forming uniform and conformal coatings that inhibit particle aggregation. This intimate coating ensures that each active particle remains electrically connected to both the electrolyte and the current collector. Recently, PANI hydrogels have been employed as binders for silicon-based LIBs [25]. In this system, PANI was polymerized in situ to form a bifunctional conformal coating on Si particles through hydrogen bonding and electrostatic interactions between positively charged PANI chains and negatively charged surface oxides on Si. The resulting Si nanoparticle–PANI composite electrodes delivered capacities ranging from 2500 to 1100 mAh g^{-1} at charge–discharge rates of 0.3–3.0 A g^{-1}, and retained approximately 91% of their capacity after 5000 cycles at a high current density of 6.0 A g^{-1}. The concept of CPG-based binder systems has further evolved into ternary composite architectures. Yu et al. [26] fabricated a three-dimensional nanostructured PPy–Si–CNT electrode that exhibited an impressive reversible capacity of ~1600 mAh g^{-1} with more than 85% capacity retention after 1000 cycles. The incorporation of CNTs significantly enhanced electron transport and electrical connectivity within the PPy framework by reinforcing both electrical and mechanical coupling between Si nanoparticles and the CPG network. Similarly, Tang et al. [27] developed another ternary system comprising active nanoparticles (Si or TiO_2), CNTs, and a PEDOT:PSS hydrogel. In this configuration, Si-based electrodes achieved a high areal capacity of 2.2 mAh cm^{-2}, while TiO_2-based electrodes delivered a capacity of 76 mAh g^{-1}. Figure 4.1 schematically illustrates the working

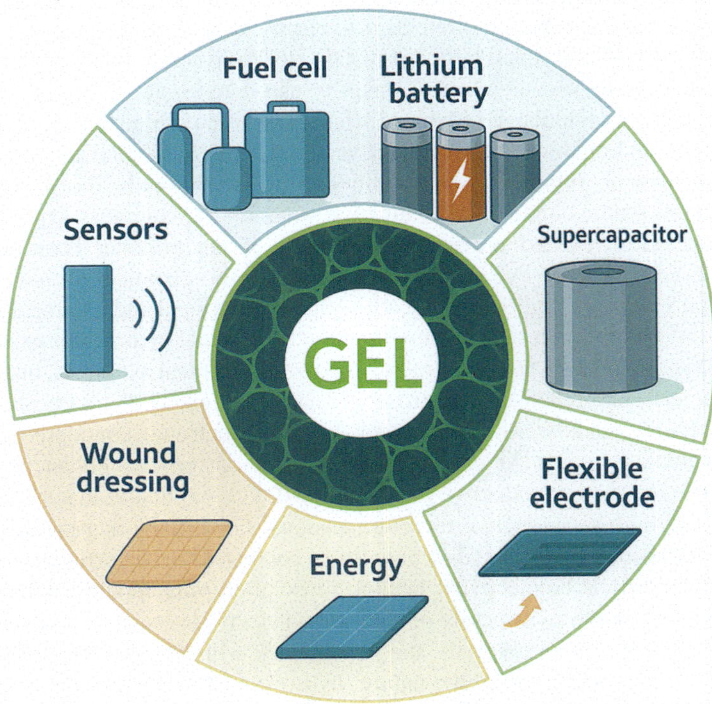

Fig. 4.1 Bioinspired conductive polymer gels for advanced energy applications

mechanisms of bioinspired conductive polymer gels across a range of energy-related applications, emphasizing their multifunctional roles in solar cells, energy storage systems, and electric vehicle technologies.

4.4 Ionically Conductive Gels

4.4.1 Hydrogel

Ion exchange-property components and a gel matrix make up ICGs. Ion conduction is made possible by the interplay of these constituents, which creates a gel structure with ion-conductive qualities. The hydrogel's pore size and distribution can be adjusted to effectively control ion diffusion and transport. Furthermore, ion selectivity and responsiveness are modulated in large part by the gels' structural design and chemical makeup. Free ions are ionized when salts (like NaCl, LiCl, KCl, $CaCl_2$, $FeCl_3$), acids (like HCl, H_2SO_4, H_3PO_4), and alkali (like NaOH, KOH) are added to the polymer solution. These free ions travel in certain directions within the gel's inner channels when an external voltage is applied, giving the gel conductivity. As the gel deforms under external forces, its resistance varies, and the gel matrix's electrostatic interaction may affect how quickly ions migrate. Furthermore, a wider variety of chemical functions are added to the gels by transition metal salts. The development of transparent devices greatly benefits from the qualities that ICGs have over ECGs, including transparency, antifreeze qualities, and self-healing capability. Liu et al. used a two-step process to create PAA-Fe-Li hydrogels [28]. Fourier-transform infrared spectroscopy showed that PAA entwined with $Li^+(H_2O)$ n and was electrostatically attached to Fe^{3+}, resulting in a synergistic improvement of the PAA-Fe-Li hydrogels'tensile strength and resilience. By increasing the ion-binding energy and chain entanglement with $Li^+(H_2O)n$, the electronegativity of PAA chains strengthened the hydrogel and gave it anti-freezing qualities. Driven by concentration differences, Li^+ diffused to the hydrogel upon contact and interacted with the interfacial water molecules. Electrical stimulation improved adhesion efficiency by increasing the diffusion rate of Li^+.Additionally, by varying the voltage intensity and duration of electrical stimulation, the adhesion strength and efficiency may be electrically programmed. In addition to increasing the hydrogel's electrical conductivity, the addition of ions decreases the water's crystallization point, increasing the hydrogel's operational temperature range. Zhang et al. [29] used a $ZnCl_2$/$CaCl_2$ binary system to create a cold-resistant cellulose hydrogel. As its water molecules entered the cellulose, $ZnCl_2$ hydrate's potent hydrogen bonding capacity sped up the disintegration of cellulose. The hydrogel was then created by adding Ca^{2+} ions as a gelling agent. While glycerol improved the hydrogel's anti-freezing qualities and maintained its flexibility at -10 °C, water effectively encouraged cellulose flocculation. Hydrogen bonds inside the hydrogel and interactions between metal ions and cellulose strands gave the system thermal reversibility. A sol–gel transition was seen during the heating–cooling cycle, which made it possible to alter the hydrogel's shape with different molds. The hydrogel's exceptional resistance to drying was facilitated by the coordination equilibrium between the metal ions and the hydroxyl groups on the cellulose strands. As a form of ion salt, ionic liquids (ILs) are extremely important in a variety of applications, such as batteries and sensors, due to their exceptional qualities, which include high conductivity, thermal stability, non-flammability, and low volatility [30]. Due to the considerable interest in the design and manufacture of IL-based gels (ionogels) using various networks, they are now widely used in tissue engineering, electrochemical biosensors, and electro-responsive drug release systems [31]. A kind of conductive hydrogel based on ILs was described by Qiu et al. [32]. It uses ion–dipole interactions and a double-network (DN) structure to provide strong stretchability and mechanical stability. The ionogel's stretchability was further enhanced by the addition of ILs without lowering its final stress. The DN structure was formed more easily by the ion–dipole interactions between the polymer network and the ionic liquid, which improved the hydrogel's mechanical and stretchability characteristics. Notably, DN ionic hydrogel demonstrated excellent stability and reproducibility in addition to good stretchability. It was a great strain sensor that could track a variety of motions, including muscle contractions and joint movements. Furthermore, by detecting deformation through variations in resistance, it showed reliable performance when used as an ion board. Fibrous materials are perfect biomaterials because they are highly adaptable and can be shaped into various shapes while maintaining their moisture permeability and breathability. In another study, Zhao et al. [33] employed 1-ethyl-3-methylimidazolium dicyanamide ([EMIM]DCA) as the solvent to fabricate an ionogel through a one-step photopolymerization process. A physically cross-linked network was formed by copolymerizing the zwitterionic monomer [2-(methacryloxy)ethyl]dimethyl-(3-sulfopropyl)ammonium hydroxide (SBMA) with acrylamide (AM), yielding P(SBMA-co-AM) ionogel fibers that inherit advantageous characteristics of fibrous materials. To enable demolding, silica tubes containing the gel were immersed in toluene [33]. The resulting ionogel fibers exhibited enhanced mechanical strength, adhesion, and antifreezing performance, which were attributed to dynamic intermolecular interactions such as hydrogen bonding, dipole–dipole interactions, and ion–dipole interactions within the network.

Ionogels can be constructed using either inorganic frameworks (e.g., SiO_2 or TiO_2) or organic matrices composed of polymers and small molecules. For instance, after fabricating photonic elastomers (PEs) incorporating SiO_2 nanoparticles, ionic liquids (ILs) were subsequently infused into the elastomer via a cosolvent infusion strategy to produce photonic ionic gels (PIGs) [34]. Hydrogen bonding played a critical role in stabilizing the photonic crystal architecture. Under external mechanical stimuli, simultaneous modulation of ionic conductivity and the lattice spacing of the SiO_2 nanoparticle array enabled synchronous transmission of optical and electrical signals, with responses correlated to the Young's modulus of the material. Furthermore, tuning the size of the SiO_2 nanoparticles allowed precise control over structural color shifts, while the type and concentration of incorporated ILs also significantly influenced the resulting optical properties.

4.4.2 Organogel

Since their introduction into lithium batteries by Feullade and Perche [35], organogel electrolytes have been among the most studied solid-state electrolytes for energy storage devices. A "polymer Li-ion battery" (PLB) employing poly(vinylidene fluoride-co-hexafluoropropylene) (PVDF-HFP) as the network was then marketed by in the 1990s [36]. An organogel is a ternary system made up of low molecular weight polar liquids, salt, and a polymer matrix. Organogel electrolytes, similar to hydrogel electrolytes, can be fabricated through either chemical or physical crosslinking. Chemically crosslinked organogels are formed via irreversible thermo- or photo-initiated polymerization, which generates permanent covalent crosslinking sites within the polymer network. In contrast, physically crosslinked organogels rely on mechanisms such as localized crystallization or polymer chain entanglement. These systems typically exhibit minor phase separation, which does not significantly affect ionic conductivity. Organic solvents play a crucial role in organogel electrolytes by reducing the crystallinity of the polymer matrix, thereby lowering the glass-transition temperature (Tg) of the resulting gel. Ideal solvents should possess low volatility, high dielectric constants, and good miscibility with the host polymer. Commonly used organic solvents include ethylene carbonate (EC), propylene carbonate (PC), diethyl carbonate (DEC), ethyl methyl carbonate (EMC), dimethyl carbonate (DMC), and dimethylformamide (DMF). In practice, mixtures of two or more solvents are often employed to enhance ionic conductivity, reduce viscosity, and broaden the electrochemical stability window. The supporting electrolyte salts typically consist of Li or Na salts with large anions and low dissociation energies, such as $LiPF_6$, $LiBF_4$, $LiCF_3SO_3$, $LiClO_4$, $NaClO_4$, and $LiN(CF_3SO_2)_2$, which facilitate efficient ion transport. The polymer host matrix is expected to provide high mechanical stability while maintaining good compatibility with the plasticizers and electrolytes. Frequently used polymer matrices include poly(vinylidene fluoride) (PVDF), poly(acrylic acid) (PAA), poly(vinyl alcohol) (PVA), poly(methyl methacrylate) (PMMA), and related polymers. Among these, PVDF has been the most extensively investigated due to several inherent advantages: (1) the presence of strongly electron-withdrawing C–F bonds that confer high electrochemical stability, particularly against high-voltage cathodes; (2) a high dielectric constant that promotes salt dissociation; (3) a relatively high Tg, which improves thermal stability; and (4) excellent processability for film fabrication. Further improvements in electrolyte uptake and reduced crystallinity can be achieved by incorporating hexafluoropropylene (HFP) units to form the PVDF-HFP copolymer, which exhibits enhanced swelling behaviour in liquid electrolytes. Conventional PVDF-based organogels are often cast as dense, nonporous films, which restrict electrolyte uptake and limit ionic conductivity. To overcome these drawbacks, fabrication techniques such as phase inversion, electrospinning, and templating have been developed to introduce controlled porosity into the polymer matrix [37]. For example, Pu et al. [38] proposed a phase inversion approach using water as a nonsolvent to produce microporous PVDF-HFP membranes. In this process, PVDF-HFP is dissolved in an acetone–water mixture and cast onto a glass substrate. Upon solvent evaporation, rapid phase separation occurs, yielding microporous membranes with porosities as high as 70–90%, which significantly enhance electrolyte uptake and ionic conductivity.

4.5 Energy Applications of Ionically Conductive Gels

4.5.1 Supercapacitors

Owing to their rapid charge discharge capability, long cycle life, and enhanced safety [39], supercapacitors typically featuring a sandwich-type configuration composed of two electrodes separated by an electrolyte have attracted considerable attention as electrochemical sensors and as promising platforms for wearable electronics, as well as for energy storage and conversion applications [40]. In the design of advanced energy storage systems, the selection of a safe, mechanically robust, and reliable electrolyte is critical, which has led to extensive research into solid-state electrolytes [41]. Among these, hydrogel electrolytes have emerged as particularly attractive solid electrolytes because of their chemically versatile networks, which can be engineered to impart multiple desirable functions, including high flexibility, excellent ionic conductivity, leak-free operation, thermal tolerance, and tunable mechanical properties. The inherently high-water content of hydrogels is especially beneficial for the development of flexible energy storage devices, as it enables the polymer network to maintain structural integrity while providing mobile ions with liquid-like transport dynamics [42]. Using a double-network architecture based on sodium alginate (SA) and polyacrylamide (PAM), Zeng et al. reported the fabrication of a hydrogel-based supercapacitor exhibiting arbitrary deformability and stable energy output [43]. In this system, potassium ferricyanide/ferrocyanide [$K_3Fe(CN)_6$/$K_4Fe(CN)_6$] redox couples and the inorganic salt Na_2SO_4 were incorporated to enhance ionic conductivity and redox activity. Meanwhile, carbon nanotubes and PEDOT: PSS were introduced to improve electrical conductivity and overall electrochemical performance. The long-chain SA polymer contributed to the formation of a three-dimensional mechanically robust support network through intermolecular interactions and polymer chain entanglement, ensuring both mechanical stability and electrochemical reliability under deformation. The charge-storage mechanism of the hydrogel-based supercapacitor is governed by the $K_3Fe(CN)_6$/$K_4Fe(CN)_6$ redox couple, which enables reversible redox reactions and establishes efficient ion-transport pathways essential for charge storage and transfer. Owing to its distinctive three-dimensional, crosslinked double-network architecture, the hydrogel supercapacitor exhibits outstanding mechanical flexibility, allowing it to undergo arbitrary deformation. This design effectively overcomes the inherently low mechanical strength and limited deformability of conventional hydrogels, which typically struggle to withstand repeated bending and twisting. The combination of high ionic conductivity in the hydrogel electrolyte and the redox-mediated charge-storage mechanism results in excellent electrochemical performance, including a high specific capacitance of 232 mF cm^{-2} at 5 mV s^{-1} and 128 mF cm^{-2} at 1 mA cm^{-2}, an energy density of 3.6 μWh cm^{-2}, and long-term cycling stability exceeding 5000 charge discharge cycles. These attributes highlight the strong potential of such systems for the development of self-powered and wearable electronic devices. In addition, Wu et al. [25] fabricated MXene hydrogel supercapacitors using a multiscale structural engineering strategy. Through unidirectional freezing, a vertically aligned, layer-by-layer architecture of MXene flakes was achieved. Compared with MXene aerogels produced by direct freeze-drying and MXene films formed by conventional layer stacking, the ordered MXene hydrogel exhibited markedly improved electrochemical performance, including an exceptional rate capability of 198 F g^{-1} at 1000 mV s^{-1} and a high capacitance of 393 F g^{-1} at 5 mV s^{-1}. The enlarged interlayer spacing within the vertically aligned MXene structure facilitated faster ion diffusion and transport, while also increasing the accessible surface area for ion adsorption. As a result, MXene-based supercapacitors with a honeycomb-like compartmental structure were obtained, delivering a high energy density of 0.1 mWh cm^{-2} and an ultrahigh areal capacitance of 2.0 F cm^{-2}. This ordered MXene hydrogel architecture provides a clear advantage over conventional layer-by-layer stacking methods and broadens the scope of energy storage applications [44]. Furthermore, Silki Sardana et al. [45] successfully developed a three-dimensional ternary nanocomposite hydrogel on carbon cloth using a two-step synthesis approach. This material functioned as a binder-free supercapacitor electrode, exhibiting an outstanding specific capacitance of 1134.28 F g^{-1} at 1 A g^{-1} and 1124 F g^{-1} at 0.25 A g^{-1}. Even at higher current densities, the electrode maintained excellent capacitance retention, preserving 82.2% at 7.5 A g^{-1} and 76.46% at 10 A g^{-1}, indicative of remarkable rate performance and cycling stability [45]. Similarly, Anil Ohlan et al. [46] reported a hierarchical three-dimensional polyaniline/doped graphene nanocomposite hydrogel supercapacitor, which achieved maximum specific power and energy densities of 0.44 kW kg^{-1} and 13.63 Wh kg^{-1}, respectively. Collectively, these studies demonstrate that composite hydrogel systems represent a scientifically robust and economically viable platform for high-performance energy storage devices, with the added advantage of being directly employed as binder-free supercapacitor electrodes.

4.6 Conclusion and Future Perspective

Gel materials represent a distinctive class of bio-inspired materials formed through the assembly of low-dimensional building blocks including 0D nanoparticles or clusters, 1D nanowires or polymer chains, and 2D nanosheets, and in some cases liquid phases—into complex, hierarchical architectures. Although the nature of the constituent blocks may vary, gel materials are unified by their interconnected three dimensional network structures, which impart unique chemical and physical properties that clearly distinguish them from isolated low-dimensional nanomaterials. The continuous gel network preserves the mechanical integrity of the entire matrix while simultaneously providing a large accessible surface area rich in active sites for reactions and anchoring. In addition, these networks offer interconnected electron-transport pathways and hierarchical porous channels, which facilitate efficient liquid infiltration and ion diffusion. Such characteristics are particularly advantageous in electrochemical systems, where interfacial reactions, volume changes induced by redox processes, and the coupled transport of electrons and ions play critical roles. Owing to these attributes, gel materials have emerged as highly attractive platforms for energy conversion and energy storage applications. To date, a wide range of gel-based nanomaterials have been developed using diverse synthesis strategies, including carbon based gels, conductive polymer gels, ionically conductive gels, and inorganic gels. These materials have demonstrated outstanding performance across multiple energy technologies, serving as electrode materials, electrolytes, self-supported current collectors, and three-dimensional binder systems in applications such as lithium-ion batteries, supercapacitors, catalysts, and fuel cells. Owing to the structural diversity of their three-dimensional architectures, a wide range of strategies have been explored to improve the performance of gel materials, including enhancements in mechanical strength, interfacial reactivity and compatibility, structural stability, and ionic/electronic conductivity. The polymer backbones of gels can be chemically engineered through several approaches, such as introducing a second network to form double-network structures, hybridizing the matrix with functional molecules or nanoparticles, or grafting specific functional groups onto the gel surface. Such chemical modifications not only reinforce the gel framework but also impart additional functionalities, including enhanced mechanical robustness, high stretchability, stimulus responsiveness, and self-healing capability. These advanced features enable the development of next-generation energy devices, such as responsive and self-healing energy conversion and storage systems. Furthermore, the mechanical properties of gels can be precisely tuned by controlling microstructural parameters, including pore size, pore distribution, and the spacing between individual building blocks. Because their mechanical behaviour is intrinsically derived from their network architecture, gel materials are particularly attractive for the fabrication of highly flexible and stretchable energy devices. In addition, their excellent processability and scalability allow gels to be manufactured using cost-effective techniques such as spray coating and inkjet printing, facilitating large-area and printable energy storage device production. Looking ahead, further progress will require more refined control over the chemical composition and physical structure of gel materials through molecular- and nanoscale-level design. This will involve the development of diverse building units, including polymer semiconductors, electrochemically active inorganic nanomaterials, and newly engineered ionic polymers, to construct advanced three-dimensional gel architectures. In contrast to conventional gel materials that are typically constructed from a single type of building block, more advanced and complex gel systems can be realized by integrating multiple components, such as carbon-based materials with conductive polymers, inorganic nanomaterials with ionic polymers, or combinations of different polymer species. By merging the complementary advantages of these distinct constituents, such hybrid systems can evolve into multifunctional gel materials that exhibit synergistic properties surpassing those of their individual components. Achieving precise control over these hybrid gels requires the rational design of new crosslinking agents and careful regulation of the physical and chemical interactions among the various building blocks. The unique attributes of gel materials—including flexibility, stretchability, potential optical transparency, and biocompatibility—enable the development of diverse and innovative energy devices. Gel materials can be directly employed as free-standing electrodes or coated onto flexible substrates to fabricate highly deformable solar cells, batteries, and supercapacitors. In addition, stimuli-responsive polymer gels that react to environmental changes such as temperature, light, or mechanical stress offer exciting opportunities for creating smart and adaptive energy devices. Beyond traditional energy technologies, gel-based systems have the potential to extend the application of energy devices into broader technological domains. For example, gel electrodes and electrolytes can be integrated into flexible biosensors, wearable electronics, and artificial skin systems, where conformability and biocompatibility are essential. Furthermore, the excellent processability and versatility of gel materials make them suitable for the fabrication of miniaturized and micro-scale energy devices. Realizing these advancements requires fundamental research supported by advanced characterization techniques to deepen understanding of gel materials. Key challenges include elucidating the relationships between mechanical properties and nanostructural features, clarifying how surface chemistry and building-block assembly influence charge transport, and uncovering the electrochemical dynamics at interfaces between gel backbones and electrolytes. Such interdisciplinary efforts will be crucial for guiding the rational design and development of next-generation gel materials with enhanced performance for energy conversion and energy storage applications.

References

1. Deligkaris K, Tadele TS, Olthuis W, van den Berg A (2010) Sens Actuator B Chem 147:765–774
2. Zhu CH, Lu Y, Chen JF, Yu SH (2014) Small 10:2796–2800
3. Ge J, Yao HB, Wang X, Ye YD, Wang JL, Wu ZY, Liu JW, Fan FJ, Gao HL, Zhang CL, Yu SH (2013) Angew Chem Int Ed 125:1698–1703
4. Cong HP, Wang P, Yu SH (2013) Chem Mater 25:3357–3362
5. Chakrabarty R, Mukherjee PS, Stang PJ (2011) Chem Rev 111:6810–6918
6. Peppas NA, Hilt JZ, Khademhosseini A, Langer R (2006) Adv Mater 18:1345–1360
7. Zhao K, Zhao Y, Xu J, Qian R, Yu Z, Ye C (2024) Stretchable, adhesive and self healing conductive hydrogels based on PEDOT:PSS-stabilized liquid metals for human motion detection. Chem Eng J 494:152971. https://doi.org/10.1016/j.cej.2024.152971
8. Zhou L, Fan L, Yi X, Zhou Z, Liu C, Fu R, Dai C, Wang Z, Chen X, Yu P et al (2018) Soft conducting polymer hydrogels cross-linked and doped by tannic acid for spinal cord injury repair. ACS Nano 12:10957–10967. https://doi.org/10.1021/acsnano.8b04609

9. Huang J, Han S, Zhu J, Wu Q, Chen H, Chen A, Zhang J, Huang B, Yang X, Guan L (2022) Mechanically stable all flexible supercapacitors with fracture and fatigue resistance under harsh temperatures. Adv Funct Mater 32:2205708. https://doi.org/10.1002/adfm.202205708
10. Guo B, Qu J, Zhao X, Zhang M (2019) Degradable conductive self-healing hydrogels based on dextran-graft-tetraaniline and N-carboxyethyl chitosan as injectable carriers for myoblast cell therapy and muscle regeneration. Acta Biomater 84:180–193. https://doi.org/10.1016/j.actbio.2018.12.008
11. Gan D, Han L, Wang M, Xing W, Xu T, Zhang H, Wang K, Fang L, Lu X (2018) Conductive and tough hydrogels based on biopolymer molecular templates for controlling in situ formation of polypyrrole nanorods. ACS Appl Electron Mater 10:36218–36228. https://doi.org/10.1021/acsami.8b10280
12. Rong Q, Lei W, Chen L, Yin Y, Zhou J, Liu M (2017) Anti-freezing, conductive self-healing organ hydrogels with stable strain-sensitivity at subzero temperatures. Angew Chem Int Ed 56:14159–14163. https://doi.org/10.1002/anie.201708614
13. Park J, Jeon N, Lee S, Choe G, Lee E, Lee JY (2022) Conductive hydrogel constructs with three-dimensionally connected graphene networks for biomedical applications. Chem Eng J 446:137344. https://doi.org/10.1016/j.cej.2022.137344
14. Qin Z, Sun X, Yu Q, Zhang H, Wu X, Yao M, Liu W, Yao F, Li J (2020) Carbon nanotubes/hydrophobically associated hydrogels as ultra stretchable, highly sensitive, stable strain, and pressure sensors. ACS Appl Mater Interfaces 12:4944–4953. https://doi.org/10.1021/acsami.9b21659
15. Zheng H, Wang S, Cheng F, He X, Liu Z, Wang W, Zhou L, Zhang Q (2021) Bioactive anti-inflammatory, antibacterial, conductive multifunctional scaffold based on MXene@CeO2 nanocomposites for infection-impaired skin multimodal therapy. Chem Eng J 424:130148. https://doi.org/10.1016/j.cej.2021.130148
16. Yu X, Qin W, Li X, Wang Y, Gu C, Chen J, Yin S (2022) Highly sensitive, weatherability strain and temperature sensors based on AgNPs@CNT composite polyvinyl hydrogel. J Mater Chem A 10:15000–15011. https://doi.org/10.1039/d2ta02559
17. Devaki SJ, Narayanan RK, Sarojam S (2014) Electrically conducting silver nanoparticle–polyacrylic acid hydrogel by in situ reduction and polymerization approach. Mater Lett 116:135–138. https://doi.org/10.1016/j.matlet.2013.10.110
18. Han L, Lu X, Wang M, Gan D, Deng W, Wang K, Fang L, Liu K, Chan CW, Tang Y, Weng LT, Yuan H (2017) A mussel-inspired conductive, self-adhesive, and self-healable tough hydrogel as cell stimulators and implantable bioelectronics. Small 13(2). https://doi.org/10.1002/smll.201601916
19. Dey K, Sandrini E, Gobetti A, Ramorino G, Lopomo NF, Tonello S, Sardini E, Sartore L (2023) Designing biomimetic conductive gelatin-chitosan-carbon black nanocomposite hydrogels for tissue engineering. Biomimetics (Basel) 8(6):473. https://doi.org/10.3390/biomimetics8060473
20. Pan L, Yu G, Zhai D, Lee HR, Zhao W, Liu N, Wang H, Tee BC-K, Shi Y, Cui Y, Bao Z (2012) Proc Natl Acad Sci U S A 109:9287–9292
21. Wang Y, Shi Y, Pan L, Ding Y, Zhao Y, Li Y, Shi Y, Yu G (2015) Nano Lett 15:7736–7741
22. Shi Y, Pan L, Liu B, Wang Y, Cui Y, Bao Z, Yu G (2014) J Mater Chem A 2:6086–6091
23. Novák P, Müller K, Santhanam KSV, Haas O (1997) Chem Rev 97:207–282
24. Kirshenbaum K, Bock DC, Lee CY, Zhong Z, Takeuchi KJ, Marschilok AC, Takeuchi ES (2015) Science 347:149–154
25. Wu H, Yu G, Pan L, Liu N, McDowell MT, Bao Z, Cui Y (2013) Nat Commun 4:1943
26. Liu B, Soares P, Checkles C, Zhao Y, Yu G (2013) Nano Lett 13:3414–3419
27. Tang Y, Zhang Y, Li W, Ma B, Chen X (2015) Chem Soc Rev 44:5926–5940
28. Liu Y, Wang P, Su X, Xu L, Tian Z, Wang H, Ji G, Huang J (2022) Electrically programmable interfacial adhesion for ultrastrong hydrogel bonding. Adv Mater 34:2108820. https://doi.org/10.1002/adma.202108820
29. Zhang XF, Ma X, Hou T, Guo K, Yin J, Wang Z, Shu L, He M, Yao J (2019) Inorganic salts induce thermally reversible and anti-freezing cellulose hydrogels. Angew Chem Int Ed 58:7366–7370. https://doi.org/10.1002/anie.201902578

30. Lei Z, Chen B, Koo YM, MacFarlane D, Introduction R (2017) Ionic liquids. Chem Rev 117:6633–6635. https://doi.org/10.1021/acs.chemrev.7b00246
31. Yang C, Suo Z (2018) Hydrogel ionotronics. Nat Rev Mater 3:125–142. https://doi.org/10.1038/s41578-018-0018
32. Qiu W, Chen G, Zhu H, Zhang Q, Zhu S (2022) Enhanced stretchability and robustness towards flexible ionotronics via double-network structure and iondipole interactions. Chem Eng J 434:134752. https://doi.org/10.1016/j.cej.2022.134752
33. Zhao L, Wang B, Mao Z, Sui X, Feng X (2022) Nonvolatile, stretchable and adhesive ionogelfiber sensor designed for extreme environments. Chem Eng J 433:133500. https://doi.org/10.1016/j.cej.2021.133500
34. Lyu Q, Wang S, Peng B, Chen X, Du S, Li M, Zhang L, Zhu J (2021) Bioinspired photonic ionogels as interactively visual ionic skin with optical and electrical synergy. Small 17:2103271. https://doi.org/10.1002/smll.202103271
35. Feuillade G, Perche P (1975) J Appl Electrochem 5:63–69
36. Tarascon JM, Gozdz AS, Schmutz C, Shokoohi F, Warren PC (1996) Solid State Ion 86–88(Part 1):49–54
37. Zhu Y, Wang F, Liu L, Xiao S, Chang Z, Wu Y (2013) Energ Environ Sci 6:618–624
38. Pu W, He X, Wang L, Jiang C, Wan C (2006) J Membr Sci 272:11–14
39. Ge M, Cao C, Biesold GM et al (2021) Recent advances in silicon-based electrodes: from fundamental research toward practical applications. Adv Mater 33(16):2004577. https://doi.org/10.1002/adma.202004577
40. Yang X, Keegan R, Gao X et al (2021) Recent advances and perspectives on thin electrolytes for high-energy-density solid-state lithium batteries. Energ Environ Sci 14:643–671. https://doi.org/10.1039/D0EE02714F
41. Zhao F, Bae J, Zhou X et al (2018) Nanostructured functional hydrogels as an emerging platform for advanced energy technologies. Adv Mater 30:1801796. https://doi.org/10.1002/adma.201801796
42. Wang Z, Li H, Tang Z et al (2018) Hydrogel electrolytes for flexible aqueous energy storage devices. Adv Funct Mater 28:1804560. https://doi.org/10.1002/adfm.20180
43. Zeng J, Dong L, Sha W et al (2020) Highly stretchable, compressible and arbitrarily deformable all-hydrogel soft supercapacitors. Chem Eng J 383:123098. https://doi.org/10.1016/j.cej.2019.123098
44. Huang X, Huang J, Yang D et al (2021) A multi-scale structural engineering strategy for high-performance MXene hydrogel supercapacitor electrode. Adv Sci 8:2101664. https://doi.org/10.1002/advs.202101664
45. Silki S, Kanika A, Sanket M et al (2022) Unveiling the surface dominated capacitive properties in flexible ternary polyaniline/NiFe2O4/reduced graphene oxide nanocomposites hydrogel electrode for supercapacitor applications. Electrochim Acta 434:141324. https://doi.org/10.1016/j.electacta.2022.141324
46. Anjli G, Silki S, Jasvir D et al (2020) Nanostructured polyaniline/graphene/Fe2O3 composites hydrogel as a high-performance flexible supercapacitor electrode material. ACS Appl Energy Mater 3:6434–6446. https://doi.org/10.1021/acsaem.0c00684

Chapter 5
Bio-Inspired Engineering of Graphene Nanocomposites for High-Performance Energy Applications

Abstract This chapter discusses recent advances in bioinspired graphene-based nanocomposites for flexible energy device applications. While graphene exhibits exceptional mechanical strength, stiffness, and electrical conductivity, the scalable assembly of graphene nanosheets into macro-sized, high-performance nanocomposites remains a significant challenge. Drawing inspiration from nacre, a benchmark natural layered material, various bioinspired assembly strategies are discussed for achieving simultaneously high strength, toughness, and electrical conductivity through synergistic structural and interfacial effects. The fundamental mechanical and electrical properties of graphene-based nanocomposites are highlighted, along with their emerging roles in flexible energy storage and harvesting devices.

Keywords Graphene nanocomposites · MXene · Supercapacitors · Lithium-ion batteries energy conversion

5.1 Introduction

Green nanotechnology has emerged as a highly valuable approach within modern science and technology due to its exceptional advantages and its ability to deliver biochemicals that are faster to produce, cost-effective, and environmentally safe. Traditionally, however, the synthesis of nanomaterials (NMs) such as metals, metal oxides, quantum dots, carbon nanotubes (CNTs), and other carbon-based nanostructures—has relied heavily on conventional chemical and physical methods. To design next-generation materials with superior electrical, optical, magnetic, and catalytic properties, it is crucial to advance innovative synthesis strategies that operate at the atomic, molecular, and macromolecular levels [1].

Among the many nanomaterials developed over the past two decades, graphene-based structures stand out as some of the most advanced and transformative. Graphene, often described as one of the most fascinating materials of the twenty-first century, consists of a single layer of sp^2-hybridized carbon atoms arranged in a

M. Y. Mir, J. A. Parray, *Nanoenergy Production*, SpringerBriefs in Energy,
https://doi.org/10.1007/978-3-032-20859-0_5

tightly packed honeycomb lattice. Despite being a relatively recent discovery, graphene composed of precisely ordered two-dimensional (2D) sheets with high crystallinity and exceptional electronic behavior has quickly proven its potential across diverse scientific and technological applications.

Since its first isolation by Novoselov and colleagues [2], this so-called "miracle material" has attracted enormous global attention, leading to a rapid expansion of research on its physics, chemistry, and applied functionalities. Graphene and its derivatives possess extraordinary characteristics, including very high thermal conductivity (around 5000 W m^{-1} K^{-1}), an impressive Young's modulus of approximately 1.0 TPa, and a remarkably large specific surface area (about 2600 m^2 g^{-1}). Furthermore, graphene demonstrates exceptional optical transparency (~97.7%), long-range ballistic electron transport at room temperature, quantum confinement effects in nanoscale ribbons, and even sensitivity sufficient for single-molecule gas detection [3]. These unique features collectively make graphene an ideal candidate for a wide spectrum of advanced applications—from energy devices and electronics to sensors, catalysis, and beyond.

The remarkable physicochemical properties of graphene have enabled its integration into a vast array of nanotechnology-based applications. These include antimicrobial coatings, flexible thin-film transistors, lithium-ion batteries, fuel cells, antibacterial and DNA-damaging activities, photocatalysis, biomedical engineering, targeted drug delivery, solar cells, photovoltaic systems, imaging technologies, P–N junction diodes, graphene-based supercapacitors, and advanced water purification systems. Graphene has also shown strong potential for the removal of organic pollutants, pharmaceutical residues, pesticides, and heavy-metal ions from aqueous environments, as well as for the development of highly sensitive chemical sensors, biosensors, and electrochemical detection platforms [4, 5].

To support such diverse and demanding applications, large quantities of high-quality graphene are needed. However, achieving scalable, cost-effective, and defect-controlled production remains a significant challenge. Over the years, several fabrication methods have been explored for synthesizing graphene from bulk graphite. These include epitaxial growth, vacuum-based thermal annealing, non-catalytic synthesis, the Scotch-tape method, micromechanical cleavage, chemical vapor deposition (CVD), thermal exfoliation, cutting of carbon nanotubes (CNTs), liquid-phase exfoliation, ultrasonication, and chemical reduction of graphene oxide [6]. Alternative and sometimes hazardous routes—such as arc discharge, chemical discharge, and CNT unzipping have also been developed for producing graphene oxide.

Each technique, however, comes with inherent limitations. For instance, although epitaxial growth can produce high-quality multilayer graphene, it cannot efficiently isolate single or bilayer conductive sheets needed for advanced energy-storage applications. Micromechanical cleavage, while effective for generating pristine graphene, suffers from low scalability due to repeated manual detachment steps, making it difficult to consistently control the number of layers obtained. These constraints emphasize the ongoing need for more reliable, scalable, and environmentally friendly synthesis approaches.

On the other hand, the cost of manufacturing thermal exfoliation is considerable. Effective exfoliation from the substrate is unclear in CVD, and layer separation is challenging to achieve without material degradation. On the other hand, the reduction of graphene oxide (GO) is thought to be a promising technique for producing graphene in large quantities. The result of this process is typically referred to as nanostructure graphene (NSG) or reduced graphene oxide (rGO). Researchers also frequently use the liquid-phase exfoliation technique as an alternative because of its ease of graphene synthesis and broad range of mixing device options. This technique involves ultrasonicating or mixing a stabilizer containing graphite to prevent the accumulation of graphene to graphite. Additionally, following a centrifugation phase, a graphene solution is the end product. Liquid-exfoliated graphene can also be applied on graphene paper using vacuum percolation, conductive ink, or nanofluids [7]. Alternatively, green technologies can be used to generate rGO/NSG from graphene oxide. First, graphite powder is used to create GO by an oxidative reaction that completely enriches the sheets with modified oxygen groups. GO is electrically insulating and causes thermal instability because to the abundance of oxygen moieties functionalized on the carbon's basal plane. As a result, the oxygen molecules are greatly moved to the p-lattice and decreased. This helps to regain electronic conductivity and create thermally established graphene. Recently, several GO reduction techniques have been described, including chemical, photocatalytic, thermal, and electrochemical techniques. Among these, electrochemical reduction is unable to address the GO precursor's intrinsic flaws. Furthermore, achieving efficient exfoliation of graphene oxide (GO) into single-layer sheets requires the GO precursor to contain a relatively low amount of oxygen. When the oxygen content is reduced, the interactions among the functional groups within GO decrease, promoting the formation of more stable structural species such as in-plane ether linkages and out-of-plane carbonyl groups. These stable oxygen-containing groups are particularly difficult to remove through electrochemical reduction techniques [3]. Among the available deoxygenation methods, thermal reduction is currently considered cost-effective. However, it involves multiple sequential steps in which water molecules and various oxygenated groups—such as hydroxyl (–OH), carboxyl (–COOH), and epoxy functionalities—are eliminated under high thermal energy, making the overall process complex. Photocatalytic reduction, on the other hand, relies on the activation of light-responsive materials under UV irradiation, enabling the gradual removal of oxygen groups. In comparison to these methods, chemical reduction is widely regarded as the simplest and most practical approach for producing ultrathin graphene sheets with large lateral dimensions and high surface area. It offers several advantages, including rapid processing, low cost, operational simplicity, and suitability for scalable production. Moreover, chemical reduction allows easier handling and provides opportunities for controlled chemical modification during or after the reduction process, making it a preferred route for the large-scale synthesis of graphene [8]. Graphene NCs have been prepared in situ using a variety of reduction processes, including chemical, thermal, bioinspired, green, and electrochemical reduction (Fig. 5.1).

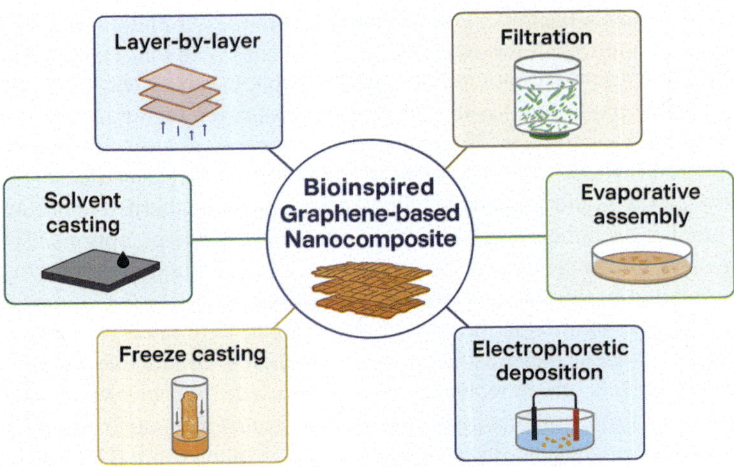

Fig. 5.1 Bioinspired assembly techniques for graphene nanocomposites

5.2 Bio-Inspired Graphene Based Nanocomposites

Several fabrication techniques such as vacuum-assisted filtration (VAF), blade coating, drop casting, and layer-by-layer (LbL) assembly—have been developed to prepare graphene and MXene based nanocomposite films, many of which draw inspiration from the highly ordered lamellar structure of natural nacre [9]. Among these methods, VAF is one of the most straightforward and efficient approaches for assembling two-dimensional nanosheets into well-structured composite films. In this process, graphene nanosheets align and stack under a controlled directional flow, gradually forming a free-standing graphene film on the filtration substrate. By simply regulating the volume or concentration of the dispersion, the final film thickness can be precisely tuned. The fabrication of mechanically robust, free-standing GO films using VAF was first reported by Dikin et al. [10], establishing the method as a foundational technique for producing graphene-based layered materials. Numerous types of graphene nanocomposite films have since been documented. However, VAF has also built the MXene nanosheets into freestanding MXene films [11]. However, the practical applicability is limited by the lengthy fabrication procedure and the small size of the resulting film. Several preparing techniques, including drop casting and blade coating, have been employed to efficiently prepare the nanocomposite films for wide area fabrication. Blade coating is typically used in continuous and scalable nanocomposite film production. Large area nanocomposite films are often made by using a blade to move and scrape slurry across the substrate. Additionally, the height and speed of the blades determine the thickness of the resulting nanocomposite layer. For example, Zhang et al. [12] used the blade coating approach to create GO films. The resulting 200 μm thick graphene films exhibit ultrahigh thermal conductivity of 1224 W/mK following graphitization and hot pressing. Furthermore, using a blade coating technique with a high shear force, Li et al. [11] created the MXene/Polyvinylidene fluoride (PVDF) nanocomposite films. Nevertheless, this technique is limited by the substance slurry's concentration and viscosity. Additionally, drop casting is frequently used to create large-scale nanocomposite films. Casting dispersion onto the substrate and then letting the solvent evaporate is how the nanocomposite films are made. Shen et al. [13], for instance, used drop casting of GO dispersion under heating to create large-area GO film. Following graphitization treatment, the high thermal conductivity graphene EMI shielding film was produced. Furthermore, the free-standing MXene film was reported by Zhong et al. [14] using a drop casting technique, which improves the adherence to the sulfur layer. Low performance, however, could be caused by the nanosheets' current aggregation during the evaporation process. Another technique for creating nanocomposite films, particularly for 2D materials, is the LbL assembly strategy. In general, the LbL assembly is a cyclical preparation method that alternately deposits two distinct materials onto the substrate to create multilayer structures. By using spin LbL, Song et al. [15] created silicone rubber/graphene nanocomposite films with a highly ordered multilayered structure. The resulting nanocomposite films have a thermal conductivity of 2.03 W/mK. But like the VAF approach, the LbL assembly process takes a lot of time.

Numerous efforts have been made in the last few years to create ultrastrong graphene nanocomposite films. Reducing voids, improving alignment degree, and fortifying interface contacts are the key components for creating high performance graphene nanocomposite films. In general, improving the mechanical characteristics of graphene nanocomposite films depends heavily on the interface contact. Synergistic interface interactions have been shown in earlier studies to be a successful method for creating high-performance graphene nanocomposite films [16]. For example, using the LbL approach, Xiong et al. [17] created transparent and ultra-robust reduced graphene oxide (rGO)/cellulose nanocrystal (CNC) nanocomposite films through ionic and hydrogen bonding interactions, significantly enhancing interfacial interactions. As a result, the resulting rGO/CNC nanocomposite film exhibits a high Young's modulus of 169 GPa and a tensile strength of 655 MPa. Additionally, the rGO/chitosan (CS)/Cu^{2+} nanocomposite films with an enhanced synergistic impact of metal-ligand coordinate bonding through copper ions were demonstrated by Cheng et al. [18]. The resulting rGO/CS/Cu^{2+} nanocomposite sheet exhibits high electrical conductivity of 23,480 S/m and tensile strength of 868.8 MPa. In the meantime, Wan et al. [19] demonstrate the new sequential bridging technique of ionic bonding and π-π interaction for creating robust and highly conductive graphene nanocomposite films. To create the sequentially bridged rGO (SBG) nanocomposite films, the GO nanosheets are first bridged with a chromic ion by ionic bonding. Next, the 1-pyrenebutyric acid N-hydroxysuccinimide ester-1-aminopyrene (PSE-AP) is added to the ionically bridged rGO hybrid. Additionally, SBG nanocomposite film exhibits tensile strength, toughness, and electrical conductivity of 821 MPa, 20 MJ/m^3, and 41,600 S/m, respectively, which are 4.0, 7.5, and 1.9 times greater than those of rGO film. Additionally, the SBG nanocomposite film shows outstanding stability in challenging conditions. High-performing graphene nanocomposite films are also produced by the other type of synergistic impact of ionic bonding and covalent contact. In order to create a chelate architecture, Wan et al. [19] added poly(dopamine) (PDA) and nickel ions to graphene oxide (GO) nanosheets. The resulting graphene nanocomposite film exhibits toughness of 19.5 MJ/m^3 and tensile strength of 417.2 MPa. Additionally, weak hydrogen bonds and strong covalent interactions can be combined to produce a synergistic effect. Using a straightforward VAF technique, Wan et al. [20] discovered strong and durable graphene nanocomposite films created by the synergistic combination of covalent and hydrogen bonding. The graphene nanocomposite film exhibits tensile strength of 526.7 MPa and toughness of 17.7 MJ/m^3 at a CS content of 5.6 weight percent, which are four and ten times more than those of natural nacre, respectively. Additionally, various synergistic effects have been studied to improve the mechanical properties of graphene nanocomposite films, including hydrogen bonding and π–π interaction [21]. The alignment and compactness of 2D nanosheets in nanocomposite films, in addition to strong interfacial contacts, are crucial for maximizing the mechanical characteristics. According to a unique technique put out by Wan et al. [22], the alignment degree of graphene nanosheets can be significantly increased by long-chain π-bonding. Ultra strong π-bridged rGO (πBG) nanocomposite films are produced by immersing the rGO films in a solution of a highly π-conjugated long-chain polymer of bis(1-pyrenemethyl) docosa-10,12-diynedioate (BPDD) and then exposing them to UV light. Furthermore, it is reported that the πBG nanocomposite film has a greater alignment than the pure rGO film, as evidenced by its lower full width at half maximum (FWHM) (24.9°) compared to 35.5°. The resulting πBG nanocomposite film is therefore 2.9 and 4.6 times stronger than pristine rGO film, respectively, with an ultrahigh strength of 1054 MPa and toughness of 36 MJ/m^3. Furthermore, the πBG nanocomposite film has an electrical conductivity of up to 119,200 S/m. Moreover, another efficient method for creating high performance nanocomposite films is to fill tiny nanosheets. Zhou et al. [23] used a filling technique involving tiny size nanosheets to demonstrate the exceptional mechanical and electrical conductivity features of MXene functionalized graphene (MrGO-AD) nanocomposites. Covalently crosslinked graphene nanosheets, which are additionally crosslinked by 1-aminopyrenedisuccinimidyl suberate (AD) molecules via the π–π interaction, are introduced to the tiny size MXene nanosheets. The gaps in nanocomposites are therefore drastically decreased, going from 15.2% for pure rGO film to 4.0% for MrGO-AD nanocomposite film for example. In contrast to the pure rGO film, which has an FWHM of 36.3°, the alignment degree of graphene nanosheets is similarly improved for MrGO-AD film, which has a narrower FWHM of 26.1°. The MrGO-AD nanocomposite sheet then exhibits enhanced electrical conductivity of 132,900 S/m, ultrahigh toughness of 42.7 MJ/m^3, and tensile strength of 699.1 MPa, respectively. In addition to Mxene nanosheets, other nanosheets like black phosphorus (BP), molybdenum sulfide (MoS_2), tungsten disulfide (WS_2), and montmorillonite (MMT) have also been used in recent years to create ultra-strong graphene nanocomposite films. Furthermore, techniques for external force-induced orientation, including freezing stretching, plasticization stretching, scanning centrifugal casting, and blade coating, are developed to decrease existing voids in nanocomposite films and increase the degree of alignment of nanosheets. Through the use of blade coating and shear alignment, Akbari et al. [24] exhibited the highly ordered and compact graphene films. Initially, the GO precursor is mixed with tiny rGO nanosheets. The extremely dense graphitic films are produced following high temperature treatment and rolling press. Because of the smaller FWHM of the Raman G band, this approach produces a higher alignment of the in-plane direction. When the rGO content is 15 weight percent, the tensile strength of the highly dense graphitic film can reach 78.5 MPa. Furthermore, in order to create highly aligned microstructure GO films, Zhang et al. [25] showed a novel chemical-structure-engineering technique for obtaining involuntarily regular stacking of GO nanosheets during the liquid-assembly process. Additionally, the produced GO film exhibits a Young's modulus of 52.3 GPa and a high strength of 445 MPa. Additionally, Li et al. [26] eliminated the wrinkles of graphene nanosheets into crystalline ordering by preparing the macroscopic graphene nanocomposite film using a continuous plasticization stretch approach. A change from a brittle to a plastic state is caused by the intercalation of solvent molecules when the direct-cast GO film is submerged in ethanol during the plastic transition phase. Additionally, the stretching procedure eliminates GO film wrinkles, yielding a high Herman's order value of 0.93. With a tensile strength of 1.1 GPa and a Young's modulus of 62.8 GPa, this technique gives graphene films exceptional mechanical properties. Additionally, Wei et al. [27] used a scanning centrifugal casting approach to demonstrate a high alignment pure graphene nanocomposite film with exceptional mechanical properties and EMI shielding performance. Additionally, the resulting 100-μm thick graphene nanocomposite film has a high tensile strength of 145 MPa and an ultrahigh EMI SE of 93 dB. Cheng's team recently used freezing stretch to align graphene nanosheets with interfacial contacts, resulting in a record-breaking graphene nanocomposite film [28]. First, a straightforward VAF approach is used to prepare the GO nanocomposite film. To create BS-GO-PCO nanocomposite films, the GO nanocomposite film is then biaxially stretched and infiltrated into the polymerization of 10,12-pentacosadiyn-1-ol (PCO, $(CH_3(CH_2)_{11}C \equiv C\text{-}C \equiv C(CH_2)_8CH_2OH)$ solution. The SB-BS-rGO nanocomposite film is then created by immersing the BS-rGO-PCO nanocomposite film in the PSE and AP solution following HI reduction. This creates π-π contact. Compared to pristine rGO films, the SB-BS-rGO nanocomposite films contain fewer voids and extremely compact graphene nanosheet stacking. Additionally, the SB-BS-rGO nanocomposite film's porosity is just 9.30%, compared to 18.7% for the pure rGO film. Furthermore, compared to the pure rGO film, the SB-BS-rGO nanocomposite film shows an ultrahigh Herman's orientation factor of 0.956. In comparison to earlier findings, the SB-BS-rGO nanocomposite film exhibits a record tensile strength of 1547 MPa. Additionally, the SB-BS-rGO nanocomposite film exhibits hardness and Young's modulus of up to 35.9 MJ/m^3 and 64.5 GPa, respectively. A possible method for creating high-performance graphene nanocomposite films is the proposed freezing stretch-induced alignment, which may also be applied to other 2D materials as MXene. Table 5.1 summarizes the major bio-inspired engineering strategies employed to design graphene nanocomposites for advanced energy conversion and storage applications.

Table 5.1 Bio-inspired engineering of graphene nanocomposites for advanced energy conversion and storage

S. no	Graphene nanocomposite	Bio-inspired design/motif	Application	Key performance metrics	References
1	Co_3O_4 or MnO_2 nanosheets integrated with graphene layered films	Nacre-like layered architecture (brick-and-mortar) to combine conductivity + mechanical integrity	Supercapacitors	High areal/volumetric capacitance; improved cyclic stability vs. plain electrode	[29]
2	Graphene-doped elastomer/ graphene-GO composite films in triboelectric layers	Skin/elastic tissue inspired hierarchical roughness and conductive networks	Flexible/stretchable TENGs (energy harvesting)	Enhanced output voltage/current and durability for wearable harvesters	[30]
3	rGO + polymeric mineral hydrogel binders; 3D interconnected graphene networks	Mussel-inspired adhesive / mineral hydrogel binders to accommodate volume change	Li-ion & Si-based battery binders / electrodes	Improved cycling stability and capacity retention for Si anodes	[31]
4	GO / graphene oxide within PVA–cellulose nanofiber hydrogels	Gradient hydrogel / biomimetic ion channels to enhance ionic transport	Zinc-ion & aqueous batteries	Higher ionic conductivity and rate capability; tunable mechanical flexibility for flexible ZIBs	[32]
5	Graphene–metal oxide (TiO_2, co_3O_4) nanocomposites with layered architectures	Plant-inspired hierarchical porosity to increase light absorption and charge separation	Photocatalysis / photoelectrochemical energy	Enhanced photocatalytic activity and charge separation efficiency; improved stability	[33]
6	Graphene-directed supramolecular polymer hydrogel membranes	Ion channel / mucilage inspired supramolecular assembly to tune transport	Membranes / ion-selective separators	Tunable ion sieving / temperature-responsive transport for energy separation devices	[34]
7	Graphene + MXene or graphene + conductive polymers nanocomposite films	Plant hierarchical architectures for multifunctionality (mechanical + electrical)	Hybrid devices & systems (multifunctional)	Combined high conductivity, flexibility, and multifunctionality (sensors + energy storage)	[35]

5.3 Energy Applications of Graphene-Based Gels

5.3.1 Supercapacitors

Supercapacitors have emerged as an important class of electrochemical energy storage devices due to their high power density, long cycling life, and ability to deliver relatively high energy densities. Carbon-based materials such as graphite, carbon nanotubes, and activated porous carbon—are widely employed as electrode materials in supercapacitors because they store charge through the formation of an electric double layer at the electrode–electrolyte interface. The specific capacitance of these carbon materials is strongly influenced by their surface area, electrical conductivity, and ability to facilitate efficient electrolyte ion diffusion. Graphene, a distinctive carbon nanomaterial, possesses exceptional electrical conductivity, mechanical robustness, flexibility, high chemical stability, and an extraordinarily large specific surface area. Owing to these properties, graphene is theoretically capable of achieving a specific capacitance as high as 550 F g^{-1} [36]. However, a major challenge arises when graphene sheets tend to restack during electrode fabrication, which significantly reduces accessible surface area and impedes ion transport. This restacking effect can be effectively minimized by assembling graphene into a three-dimensional (3D) hydrogel network. The resulting porous architecture provides multidirectional pathways for electron movement, improves electrolyte accessibility, and shortens ion diffusion distances within the electrode, ultimately enhancing capacitance performance. To address this issue, Zhang and Shi [36] developed a self-assembled graphene hydrogel using a one-step hydrothermal treatment followed by chemical reduction with hydrazine or hydroiodic acid. The resulting 3D porous graphene material served as an efficient supercapacitor electrode, exhibiting a high specific capacitance of 220 F g^{-1} at a current density of 1 A g^{-1}. Remarkably, it also retained a capacitance of 165 F g^{-1} in aqueous electrolyte even at an ultrafast discharge rate of 100 A g^{-1}. These enhanced electrochemical characteristics were attributed to the modified morphology and chemical structure of the graphene hydrogel, which provided abundant active sites and improved ion/electron transport pathways. The graphene sheets' conductivity was further enhanced by the chemical reduction, which eliminated any remaining oxygenated groups. The unique three-dimensional (3D) architecture of the graphene hydrogel exposed a large fraction of graphene sheets to the electrolyte, while its interconnected porous channels facilitated efficient ion transport throughout the material. Notably, this graphene-hydrogel-based electrode was fabricated without polymer binders or conductive additives, simplifying the processing steps and eliminating potential performance losses caused by extraneous materials. Owing to their remarkable mechanical strength and inherent flexibility, graphene hydrogels are excellent candidates for flexible solid-state supercapacitors. For example, Xu et al. [37] developed a solid-state supercapacitor using graphene hydrogel films as electrodes combined with an H_2SO_4–polyvinyl alcohol (PVA) gel electrolyte. Using a graphene hydrogel thin film approximately 120 μm in thickness, the device exhibited outstanding mechanical flexibility, long-term cycling stability, and a high gravimetric specific capacitance of 186 F g^{-1}, along with an impressive areal capacitance of 372 mF cm^{-2}. To further enhance the electrochemical performance of graphene-based hydrogels as supercapacitor electrodes, several engineering strategies have been explored. One effective approach involves modifying the surfaces of graphene sheets within the hydrogel network. Xu et al. [38] functionalized graphene hydrogels by chemically reducing graphene oxide (GO) using hydroquinone, which served simultaneously as a reducing agent and functionalizing molecule. The exceptionally large surface area of the graphene hydrogel allowed it to accommodate a substantial quantity of hydroquinone molecules, significantly boosting pseudocapacitive contributions. Because these hydroquinone molecules were directly π-π bonded to the graphene sheets, rapid electron transfer between the graphene substrate and hydroquinone was facilitated, enabling fast Faradaic reactions and maximizing the pseudocapacitive effect. As a result, the functionalized graphene hydrogel electrodes achieved a specific capacitance of 441 F g^{-1}. In a subsequent study, the same research group further enhanced ion accessibility by introducing H_2O_2 during the self-assembly of graphene sheets, thereby producing a holey graphene framework [39]. The H_2O_2 selectively etched carbon atoms around defect sites in GO, generating carbon vacancies that gradually evolved into nanopores distributed across the graphene sheets. This holey hydrogel structure exhibited superior mechanical robustness and an exceptionally high specific surface area of 1560 m^2 g^{-1}. Importantly, its enlarged interconnected pores enabled efficient ion transport even under severe mechanical compression. A fully packaged supercapacitor constructed from these holey graphene hydrogels delivered gravimetric and volumetric energy densities of 35 Wh kg^{-1} and 49 Wh L^{-1}, respectively values comparable to those of commercial lead-acid batteries. Doping graphene gel frameworks with nitrogen, boron, sulphur, or other atoms is another technique for creating high-performance electrode materials. Using GO and organic amines as precursors, Chen et al. [40] created nitrogen-doped graphene hydrogels. The pore size and other microstructural characteristics of the graphene hydrogels may be adjusted by varying the species and concentrations of the organic amines. Because of the hanging chains of the organic amines outside the plane and the creation of hydrogen bonds, N doping in the carbon plane may increase separation and prevent stacking between graphene sheets, increasing electrochemical performance. With an ultrafast charge/discharge rate of 185.0 Ag^{-1}, the resultant supercapacitors could achieve a highpower density of 205.0 $kWkg^{-1}$. Using ammonia boron trifluoride (NH_3BF_3) as the precursor through a hydrothermal reaction, Wu et al. [41] co-doped graphene aerogels with nitrogen and boron. The graphene framework's electrical conductivity was improved, and a high power density of about 1600 Wkg^{-1} was attained. To further utilize the promise of graphene-based gels for electrochemical applications, new methods for device structure design and microstructural engineering of graphene gels have been proposed in addition to chemical modification. Using an aqueous GO dispersion and arbitrarily shaped Zn particles, Maiti et al. [42] created a straightforward and adaptable graphene gelation technique that can be used to shape engineer micrometer-thick hydrogels. At the Zn surfaces, hydrogel films formed spontaneously. This site-specific gelation allowed for a great deal of control since the area could be adjusted by changing the Zn film size and the thickness could be adjusted by changing the immersion period. The resultant graphene-gel-based supercapacitors demonstrated a high areal power density of 369.8 $mWcm^{-2}$ and an energy density of up to 2.73 Wh cm^{-2}. In order to get greater performance, graphene hydrogels have also been utilized to create asymmetric supercapacitors in addition to just symmetrical ones. By employing vertically aligned MnO_2 nanoplates on nickel foam as the positive electrode and graphene hydrogel as the negative electrode, Gao et al. [43] created an asymmetric supercapacitor. At the Zn surfaces, hydrogel films formed spontaneously. This site-specific gelation allowed for a great deal of control since the area could be adjusted by changing the Zn film size and the thickness could be adjusted by changing the immersion period. The resultant graphene-gel-based supercapacitors demonstrated a high areal power density of 369.8 $mWcm^{-2}$ and an energy density of up to 2.73 Wh cm^{-2}. In order to get greater performance, graphene hydrogels have also been utilized to create asymmetric supercapacitors in addition to just symmetrical ones. By employing vertically aligned MnO_2 nanoplates on nickel foam as the positive electrode and graphene hydrogel as the negative electrode, Gao et al. [43] created an asymmetric supercapacitor. The potential windows of the two electrodes were complimentary. The resulting supercapacitor demonstrated an energy density of 23.2 Wh kg^{-1} and a power density of 1.0 $kWkg^{-1}$, and it could be cycled across a broad potential range of 0–2.0 V.

5.4 Energy Conversion

Fundamental electrochemical reactions including the hydrogen evolution reaction (HER), oxygen evolution reaction (OER), and oxygen reduction reaction (ORR) play a central role in a wide range of energy conversion and storage technologies such as fuel cells, solar cells, and metal–air batteries. Noble metals like Pt, Ru, and Ir are known to exhibit exceptional catalytic performance for these reactions; however, their limited availability and high cost significantly restrict their widespread practical application. As a result, graphene and graphene-based hybrid catalysts have recently emerged as promising alternatives due to their high catalytic activity, low cost, and the absence of precious metals. Graphene-based gel materials, in particular, hold great potential for enhancing catalytic properties because of their unique structural and chemical features. These self-supporting gels possess strong mechanical integrity, allowing them to function directly as working electrodes while maintaining excellent durability. Their highly hydrophilic, interconnected frameworks improve electrode wettability and facilitate rapid electrolyte penetration, which enhances reaction kinetics and overall catalytic efficiency. Moreover, the porous gel network enables efficient gas transport within the electrode structure an essential factor for gas-involving reactions such as HER, OER, and ORR. A growing number of studies have demonstrated the effectiveness of graphene-gel-based catalysts, highlighting their potential as next-generation materials for high-performance electrochemical energy systems. Chen et al. [44] developed an innovative N,O–dual-doped graphene CNT hydrogel electrocatalyst by assembling chemically transformed graphene (containing inherent oxygen functionalities) with carbon nanotubes through a simple filtration process, followed by ammonia-assisted nitrogen doping. This self-supporting hydrogel displayed exceptional durability in both alkaline and strongly acidic environments and exhibited superior OER performance compared to the noble metal catalyst IrO_2 as well as several transition-metal-based catalysts. The enhanced catalytic kinetics were attributed to the material's unique structural characteristics, including its three-dimensional interconnected framework, well-developed porosity, and dual N/O doping, which introduced abundant active catalytic sites (such as C–N and C–O–C configurations) for the OER. Graphene-gel-based composites incorporating additional active particles have also been explored as hybrid electrocatalysts for various electrochemical reactions. For instance, Chen et al. [45] prepared N-doped graphene hydrogels decorated with NiCo particles for the OER and MoS_x species for the HER. These hybrid materials leveraged several advantageous structural features: a highly hydrated and ion-permeable framework, a conductive 3D network, strong mechanical integrity, and abundant active catalytic centers provided by the incorporated species. As a result, both catalysts demonstrated significantly improved catalytic activity and long-term stability. The MoS_x–graphene hydrogel delivered an HER overpotential of 140.6 mV, while the NiCo–graphene hydrogel showed a low OER over potential of 350 mV and a high catalytic current density of 145.3 mA cm^{-2}. In another study, an N-doped graphene aerogel embedded with Fe_3O_4 nanoparticles served as an efficient cathode catalyst for the ORR. Compared with Fe_3O_4 nanoparticles supported on N-doped carbon black or N-doped graphene sheets, the gel-based catalyst exhibited a more positive onset potential, enhanced cathodic current density, reduced H_2O_2 production, and a higher electron transfer number in alkaline media [46]. These results highlight the crucial role of the graphene-gel support specifically its 3D porous network and large specific surface area in achieving improved ORR performance.

5.5 Conclusion and Future Perspective

We discussed several instances of bio-inspired and bio-derived graphene nanosystems in this featured mini-review, along with their uses in a range of cutting-edge technical domains. These nanoplatforms have demonstrated the improved qualities required to create systems with improved performance, such as graphene‘s ability to absorb light, which is inspired by leaf photosynthesis; adhesion and self-cleaning properties, which are inspired by geckos' feet; eel's ability to generate electricity; super hydrophobicity; and a variety of sensing properties, which are inspired by human sensory organs. However, in the realm of biomass-derived nanomaterials, graphene is still relatively new. Numerous graphene forms and functions are taken into account in bio-inspiration, which makes it appropriate for creating novel nanosystems for energy and healthcare purposes. Graphene's exceptional electrical conductivity is essential for creating flexible and sensitive sensors. The interaction properties (covalent bonding, π–π staking, etc.) and biocompatibility result in medication delivery, gene delivery, bioimaging, and disease treatment. These nanosystems are a greener chemical technique that can enhance the performance of biomedical devices based on bioinspired graphene and are appropriate for bioactives (cells, DNA, RNA, antibodies, etc.). Furthermore, the authors anticipate that future research will focus on promoting bio-inspired graphene nanosystems for the biomedical field and exploring scaling up techniques due to its affordability and bio acceptability.

References

1. Rao CEE, Sood AE, Subrahmanyam KE, Govindaraj A (2009) Angew Chem Int Ed 48(42):7752–7777
2. Novoselov KS, Geim AK, Morozov SV, Jiang D, Zhang Y, Dubonos SV, Grigorieva IV, Firsov AA (2004) Science 306:666–669
3. Potbhare K, Bagade R, Chouke PB, Zahra S, Lambat T, Bagade MB, Chaudhary RG (2019) Mater Today Proc 15:454–463
4. Zhu L, Liu Z, Xia P, Li H, Xie Y (2018) Ceram Int 44:849–856
5. Potbhare AK, Chouke PB, Mondal A, Thakare R, Mondal S, Chaudhary RG, Rai AR (2020) Mater Today Proc 29:939–945
6. Kwak J, Kim SY, Jo Y, Kim NY, Kim SY, Lee Z, Kwon SY (2018) Adv Mater 30:1800022
7. Abdullah AH, Ismail Z, Abidin ASZ, Yusoh K (2019) Mater Chem Phys 222:11–19
8. Zhang D, Zhang Y, Luo Y, Zhang Y, Li X, Yu X, Dind X, Chu PK, Sun L (2018) High-performance asymmetrical supercapacitor composed of rGO-enveloped nickel phosphite hollow spheres and N/S co-doped rGO aerogel. Nano Res 11:1651–1663. https://doi.org/10.1007/s12274-017-1780-3
9. Dikin DA, Stankovich S, Zimney EJ, Piner RD, Dommett GHB, Evmenenko G, Nguyen ST, Ruoff RS (2007) Preparation and characterization of graphene oxide paper. Nature 448:457–460. https://doi.org/10.1038/nature06016
10. Lukatskaya MR, Mashtalir O, Ren CE, Dall'Agnese Y, Rozier P, Taberna PL, Naguib M, Simon P, Barsoum MW, Gogotsi Y (2013) Cation intercalation and high volumetric capacitance of two-dimensional titanium carbide. Science 341:1502–1505. https://doi.org/10.1126/science.1241488
11. Li Y, Zhou B, Shen Y, He C, Wang B, Liu C, Feng Y, Shen C (2021) Scalable manufacturing of flexible, durable Ti3C2Tx MXene/Polyvinylidene fluoride film for multifunctional electromagnetic interference shielding and electro/photothermal conversion applications. Compos Part B Eng 217:108902. https://doi.org/10.1016/j.compositesb.2021.108902
12. Zhang X, Guo Y, Liu Y, Li Z, Fang W, Peng L, Zhou J, Xu Z, Gao C (2020) Ultrathick and highly thermally conductive graphene films by self-fusion. Carbon 167:249–255. https://doi.org/10.1016/j.carbon.2020.05.051
13. Shen B, Zhai W, Zheng W (2014) Ultrathin flexible graphene film: an excellent thermal conducting material with efficient EMI shielding. Adv Funct Mater 24:4542–4548. https://doi.org/10.1002/adfm.201400079
14. Zhong X, Wang D, Sheng J, Han Z, Sun C, Tan J, Gao R, Lv W, Xu X, Wei G, Zou X, Zhou G (2022) Freestanding and sandwich MXene-based cathode with suppressed lithium

polysulfides shuttle for flexible lithium-sulfur batteries. Nano Lett 22:1207–1216. https://doi.org/10.1021/acs.nanolett.1c04377

15. Song J, Chen C, Zhang Y (2018) High thermal conductivity and stretchability of layer-by-layer assembled silicone rubber/graphene nanosheets multilayered films. Compos Part A Appl Sci Manuf 105:1–8. https://doi.org/10.1016/j.compositesa.2017.11.001
16. Gong S, Zhang Q, Wang R, Jiang L, Cheng Q (2017) Synergistically toughening nacrelike graphene nanocomposites via gel-film transformation. J Mater Chem A 5:16386–16392. https://doi.org/10.1039/c7ta03535g
17. Xiong R, Hu K, Grant AM, Ma R, Xu W, Lu C, Zhang X, Tsukruk VV (2016) Ultrarobust transparent cellulose nanocrystal-graphene membranes with high electrical conductivity. Adv Mater 28:1501–1509. https://doi.org/10.1002/adma.201504438
18. Cheng Y, Peng J, Xu H, Cheng Q (2018) Glycera-inspired synergistic interfacial interactions for constructing ultrastrong graphene-based nanocomposites. Adv Funct Mater 28:1800924. https://doi.org/10.1002/adfm.201800924
19. Wan S, Xu F, Jiang L, Cheng Q (2017) Superior fatigue resistant bioinspired graphenebased nanocomposite via synergistic interfacial interactions. Adv Funct Mater 27:1605636. https://doi.org/10.1002/adfm.201605636
20. Wan S, Peng J, Li Y, Hu H, Jiang L, Cheng Q (2015) Use of synergistic interactions to fabricate strong, tough, and conductive artificial nacre based on graphene oxide and chitosan. ACS Nano 9:9830–9836. https://doi.org/10.1021/acsnano.5b02902
21. Song P, Xu Z, Wu Y, Cheng Q, Guo Q, Wang H (2017) Super-tough artificial nacre based on graphene oxide via synergistic interface interactions of pi-pi stacking and hydrogen bonding. Carbon 111:807–812. https://doi.org/10.1016/j.carbon.2016.10.067
22. Wan S, Chen Y, Wang Y, Li G, Wang G, Liu L, Zhang J, Liu Y, Xu Z, Tomsia AP, Jiang L, Cheng Q (2019) Ultrastrong graphene films via long-chain πbridging. Matter 1:389–401. https://doi.org/10.1016/j.matt.2019.04.006
23. Zhou T, Wu C, Wang Y, Tomsia AP, Li M, Saiz E, Fang S, Baughman RH, Jiang L, Cheng Q (2020) Super-tough MXene-functionalized graphene sheets. Nat Commun 11:2077. https://doi.org/10.1038/s41467-020-15991-6
24. Akbari A, Cunning BV, Joshi SR, Wang C, Camacho-Mojica DC, Chatterjee S, Modepalli V, Cahoon C, Bielawski CW, Bakharev P, Kim G-H, Ruoff RS (2020) Highly ordered and dense thermally conductive graphitic films from a graphene oxide/reduced graphene oxide mixture. Matter 2:1198–1206. https://doi.org/10.1016/j.matt.2020.02.014
25. Zhang Y, Wang S, Tang P, Zhao Z, Xu Z, Yu ZZ, Zhang HB (2022) Realizing spontaneously regular stacking of pristine graphene oxide by a chemicalstructure-engineering strategy for mechanically strong macroscopic films. ACS Nano 16:8869–8880. https://doi.org/10.1021/acsnano.1c10561
26. Li P, Yang M, Liu Y, Qin H, Liu J, Xu Z, Liu Y, Meng F, Lin J, Wang F, Gao C (2020) Continuous crystalline graphene papers with gigapascal strength by intercalation modulated plasticization. Nat Commun 11:2645. https://doi.org/10.1038/s41467-020-16494-0
27. Wei Q, Pei S, Qian X, Liu H, Liu Z, Zhang W, Zhou T, Zhang Z, Zhang X, Cheng HM, Ren W (2020) Superhigh electromagnetic interference shielding of ultrathin aligned pristine graphene nanosheets film. Adv Mater 32:1907411. https://doi.org/10.1002/adma.201907411
28. Wan S, Chen Y, Fang S, Wang S, Xu Z, Jiang L, Baughman RH, Cheng Q (2021) High-strength scalable graphene sheets by freezing stretch-induced alignment. Nat Mater 20:624–631. https://doi.org/10.1038/s41563-020-00892-2
29. He C, Liang Y, Gao P, Cheng L, Shi D, Xie X, Li RKY, Yang Y (2017) Bio-inspired Co3O4/graphene layered composite films as self-supported electrodes for supercapacitors. Compos Part B Eng 121:68–74
30. Xie B, Guo Y, Chen Y, Zhang H, Xiao J, Hou M, Liu H, Ma L, Chen X, Wong C (2024) Advances in graphene-based electrode for triboelectric nanogenerator. Nanomicro Lett 17(1):17. https://doi.org/10.1007/s40820-024-01530-1

31. Tian M, Chen X, Sun S, Yang D, Wu P (2019) A bioinspired high-modulus mineral hydrogel binder for improving the cycling stability of microsized silicon particle-based lithium-ion battery. Nano Res 12(5):1121–1127
32. Wang Q, Huang J, Qi L, Li M, Wang S, Chen J, Sui Z, Bi T, Tang Q, Yu L, Hu P (2025) A bioinspired gradient hydrogel electrolyte network with optimized interfacial chemistry toward robust aqueous zinc-ion batteries. ACS Nano 19(29):26770–26781
33. Potbhare AK, Aziz ST, Ayyub MM, Kahate A, Madankar R, Wankar S, Dutta A, Abdala A, Mohmood SH, Adhikari R, Chaudhary RG (2024) Bioinspired graphene-based metal oxide nanocomposites for photocatalytic and electrochemical performances: an updated review. Nanoscale Adv 6(10):2539–2568
34. Liu J, Wang N, Yu LJ, Karton A, Li W, Zhang W, Guo F, Hou L, Cheng Q, Jiang L, Weitz DA (2017) Bioinspired graphene membrane with temperature tunable channels for water gating and molecular separation. Nat Commun 8(1):2011
35. Li L, Cheng Q (2022) Bioinspired nanocomposite films with graphene and MXene. Giant 12:100117
36. Zhang L, Shi G (2011) J Phys Chem C 115:17206–17212
37. Xu Y, Lin Z, Huang X, Liu Y, Huang Y, Duan X (2013) ACS Nano 7:4042–4049
38. Xu Y, Lin Z, Huang X, Wang Y, Huang Y, Duan X (2013b) Adv Mater 25:5779–5784
39. Xu Y, Lin Z, Zhong X, Huang X, Weiss NO, Huang Y, Duan X (2014) Nat Commun 5:4554
40. Chen P, Yang JJ, Li SS, Wang Z, Xiao TY, Qian YH, Yu SH (2013) Nano Energy 2:249–256
41. Wu ZS, Winter A, Chen L, Sun Y, Turchanin A, Feng X, Müllen K (2012) Adv Mater 24:5130–5135
42. Maiti UN, Lim J, Lee KE, Lee WJ, Kim SO (2014) Adv Mater 26:615–619
43. Gao H, Xiao F, Ching CB, Duan H (2012) ACS Appl Mater Interfaces 4:2801–2810
44. Chen S, Duan J, Jaroniec M, Qiao S-Z (2014) Adv Mater 26:2925–2930
45. Chen S, Duan J, Tang Y, Jin B, Qiao SZ (2015) Nano Energy 11:11–18
46. Giri G, Verploegen E, Mannsfeld SCB, Atahan-Evrenk S, Kim DH, Lee SY, Becerril HA, Aspuru-Guzik A, Toney MF, Bao Z (2011) Nature 480:504–508

Chapter 6
Bio-Inspired Nanomaterial Technologies for Integrated Self-Powered Systems: From Smart Wearables to Off-Grid Rural Applications

Abstract Research on bio-inspired nanomaterial technologies for self-powered systems has intensified due to the rising need for sustainable, independent energy sources. Bio-inspired nanostructures have developed as a ground-breaking method for effectively harvesting, storing, and controlling energy by mimicking the structural, functional, and adaptive strategies of natural materials. In order to create self-sustaining systems that don't require external power sources, this chapter examines the design concepts and integration techniques of bio-derived and biomimetic nanomaterials. The focus is on hybrid designs that combine photovoltaic, piezoelectric, and triboelectric processes to provide multifunctional energy platforms appropriate for off-grid rural applications and wearable electronics. System integration frameworks, manufacturing difficulties, and the financial and environmental effects of implementing bio-inspired energy systems are also covered in this chapter. In the end, it emphasizes how nature-driven nanotechnology might revolutionize the development of decentralized rural micro energy infrastructures and self-powered smart wearables that are in line with global sustainability objectives.

Keywords System integration · Self-powered systems · Electronic tongue · Electronic skin · Off-grid rural power units

6.1 Introduction

A revolutionary area of nanotechnology is represented by bio-inspired nanoscale materials, which take inspiration from biological processes to produce cutting-edge materials with remarkable qualities. These materials have a wide range of applications in fields including biomedical engineering, renewable energy, and environmental sustainability since they are made to mimic natural molecular processes. The hardness of nacre, the hydrophobicity of lotus leaves, and the strength of spider silk

M. Y. Mir, J. A. Parray, *Nanoenergy Production*, SpringerBriefs in Energy,
https://doi.org/10.1007/978-3-032-20859-0_6

are examples of natural designs that may be used as models to create nanoscale systems with exceptional flexibility, durability, and multifunctionality. One of the main benefits of bio-inspired materials is their ability to self-assemble like biological molecules, which makes it possible to precisely arrange nanostructures for applications including drug delivery systems, tissue scaffolds, and biosensors. Additionally, a lot of these materials are naturally environmentally beneficial since they use green synthesis methods or biodegradable components that reduce toxic consequences, supporting the ideas of the circular economy and lowering environmental footprints. These developments provide state-of-the-art technology solutions and sustainable substitutes that mimic the effectiveness and flexibility of natural systems [1]. When compared to traditional technologies, the integration of bio-inspired nanomaterials and devices has made it possible to significantly improve energy conversion, storage, and harvesting capacities. Bio-inspired solar cells, biofuel cells, and hybrid energy harvesters have been inspired by biological energy processes as photosynthesis, metabolism, and energy transduction [2]. Researchers are creating lightweight, adaptable, and scalable energy systems that may be used in a variety of settings and applications by implementing natural design principles. By promoting the use of plentiful, non-toxic materials and lowering reliance on rare or hazardous components, these tactics promote the development of sustainable energy. Utilizing bio-inspired methods in renewable energy systems reduces energy loss, improves resource efficiency, and lessens environmental effects. Additionally, these materials open up new ways to gather energy from non-traditional sources including ambient motion, vibrations, and waste heat, increasing the potential for renewable energy [3]. Because this discipline is multidisciplinary, biologists, chemists, materials scientists, and engineers may work together to drive innovation and accelerate developments in sustainable energy. Recent developments emphasize how organic waste materials, such as biomass, shells, and agricultural leftovers, and biofluids, such as blood, sweat, and saliva, may be used to create sustainable nanostructures. These bioresources, which are abundant in proteins, polysaccharides, and enzymes, serve as organic precursors or templates for functional nanomaterials that have energy-harvesting, catalytic, or electrochemical properties. In addition to creating biocompatible, biodegradable, and low-toxicity materials, using such feedstocks is consistent with circular economy frameworks and green chemistry concepts. Researchers are creating new generations of multipurpose materials for wearable electronics, self-powered systems, and environmentally friendly energy devices by mimicking biological processes and integrating waste valorization (Fig. 6.1).

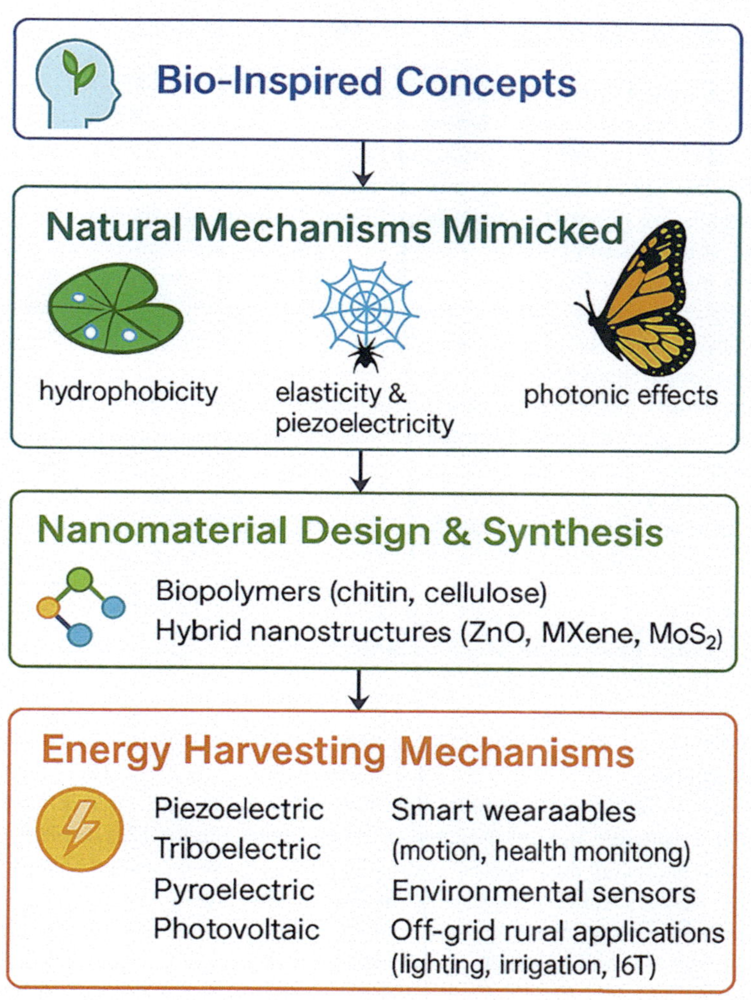

Fig. 6.1 Bio-inspired nanomaterial technologies for integrated self powered systems

6.2 Design Principles and Mechanisms

Because of their intricate multiscale and hierarchical architectures, biological systems present a diverse range of materials and structural designs with exceptional functional properties. Shaped and optimized through billions of years of evolution, these natural structures offer valuable inspiration for the development of advanced materials capable of addressing critical challenges in energy storage, energy harvesting, and overall energy efficiency [4]. This section explores the fundamental design principles and underlying natural mechanisms that contribute to the outstanding performance of bio-inspired materials, highlighting how lessons from nature can be translated into innovative engineering solutions.

6.2.1 *Hierarchical and Photonic Designs in Nature*

6.2.1.1 Nacre (Mother of Pearl)

Nacre is distinguished by its outstanding mechanical strength and toughness, which arise from its natural "brick-and-mortar" architecture. In this structure, stiff aragonite platelets are bonded together by compliant organic polymer layers, enabling efficient stress distribution and resistance to fracture under mechanical loading. Drawing inspiration from this design, researchers have developed nacre-mimetic composite materials for energy storage applications, particularly in lithium-ion batteries. The ordered, layered architecture of these bio-inspired materials enhances both mechanical robustness and ion transport pathways, leading to notable improvements in battery performance. Recent studies report that nacre-inspired electrodes composed of graphene oxide integrated with biopolymer matrices can deliver specific capacities of up to 500 mAh g^{-1}, nearly double that of conventional graphite electrodes, which typically exhibit capacities around 250 mAh g^{-1} [5]. Moreover, these electrodes demonstrate excellent long-term durability, retaining over 90% of their initial capacity even after 1000 charge–discharge cycles [6].

6.2.1.2 Lotus Leaves and Morpho Butterfly Wings: Light Trapping Technologies

Water contact angles exceeding 160° are characteristic of the lotus leaf's renowned superhydrophobic behaviour. This effect originates from its hierarchical micro- and nano-scale surface textures, which impart strong self-cleaning capabilities that are particularly valuable in contaminated or dusty environments [7]. Inspired by this natural design, recent advances in coating technologies have enabled solar panels to achieve efficiency improvements of up to 35% in dust-prone conditions by substantially reducing dirt adhesion and minimizing maintenance requirements. The integration of micro- and nano-structured surface features with a thin wax-like layer creates a highly water-repellent surface that promotes efficient water droplet rolling, facilitating a self-cleaning process that prevents particle accumulation and helps preserve photovoltaic performance over time [8]. Similarly, the iridescent appearance of Morpho butterfly wings arises from their distinctive photonic nanoarchitectures, which manipulate light through interference and diffraction rather than pigmentation. These naturally occurring photonic structures have inspired the development of advanced light-management coatings for photovoltaic devices. Recent innovations based on such bio-inspired designs have demonstrated efficiency enhancements of up to 20% under variable illumination conditions, making them particularly advantageous for solar energy harvesting in low-light or fluctuating environments [9].

6.2.2 *Mechanisms Underpinning Enhanced Performance*

Through millions of years of evolution, nature has developed highly optimized structures that provide ideal templates for advanced material design. By emulating these biological architectures, bio-inspired materials can achieve enhanced charge mobility, minimized optical losses, and improved insulation performance. This section emphasizes the relevance of key natural mechanisms to modern energy technologies, including efficient ion transport within layered and hierarchical structures, light control enabled by nano-optical features, and bio-mimetic adhesion strategies. Together, these principles play a crucial role in advancing energy storage, energy harvesting, and overall system efficiency [10].

6.2.2.1 Energy Storage

Layered architectures inspired by the "brick-and-mortar" structure of nacre markedly enhance both ionic transport and mechanical robustness in electrode materials. By replicating the hierarchical stratification characteristic of mollusc shells, these designs commonly employ stacked graphene oxide layers reinforced with biopolymeric matrices, which together provide structural integrity and efficient ion pathways [11]. Consequently, such bio-inspired electrodes exhibit outstanding electrochemical performance, including exceptional cycling stability with more than 95% capacity retention after 1000 charge–discharge cycles and substantially higher specific capacities of up to 450 mAh g^{-1} nearly double that of conventional graphite electrodes, which typically deliver around 250 mAh g^{-1}. These performance gains are particularly important for electric vehicles and grid-scale energy storage systems, where fast charging capability and long-term operational reliability are essential [12].

6.2.2.2 Energy Harvesting

Biomimetic designs inspired by the light-manipulating wings of the Morpho butterfly enable the effective integration of complex photonic functionalities into photovoltaic coatings. The nano-chitin-based layered architectures found in these wings interact with incident light through diffraction and interference mechanisms, leading to a 15–20% enhancement in solar absorption, particularly at oblique angles and under diffuse illumination conditions. Compared with conventional anti-reflective coatings, these bio-inspired photonic structures sustain higher energy yields across a wide range of lighting environments, making them especially advantageous for photovoltaic deployment in urban, shaded, or variable-sunlight settings.

6.2.2.3 Energy Efficiency

Lotus-leaf-inspired surfaces exhibit water contact angles exceeding 150°, imparting strong superhydrophobic and self-cleaning properties that effectively shield solar panels from moisture accumulation and dust deposition. Empirical studies show that photovoltaic modules treated with such coatings can retain up to 95% of their efficiency under dry conditions, whereas untreated panels subjected to soiling effects often experience efficiency losses, declining to nearly 70%. Compared with conventional fluorinated coatings, these advanced bio-inspired surfaces offer approximately 30% greater resistance to UV degradation and reduce maintenance requirements by more than 60%, making them particularly suitable for long-term outdoor deployment. At the same time, adhesives inspired by the attachment mechanism of gecko feet based on densely packed nanostructures that exploit van der Waals forces enable strong, reversible bonding without leaving chemical residues. Relative to traditional polyurethane or silicone-based sealants, these bio-mimetic adhesives enhance air-tightness and thermal sealing performance in energy infrastructure by roughly 20–30%. Table 6.1 presents a consolidated overview of state-of-the-art bio-inspired materials, the underlying biological mechanisms, and their corresponding energy-related applications, while also outlining key performance improvements alongside considerations of commercial viability and scalability.

Table 6.1 Summary of bio-inspired material mechanisms, applications, and performance

S. no	Bio-inspired material	Mechanism	Application	Performance benefit	References
1	Nacre-inspired electrodes	Ion transport	Lithium-ion batteries	Specific capacities in the range of 450–500 mAh g^{-1}, together with capacity retention exceeding 95% after 1000 charge–discharge cycles	[13]
2	Lotus-leaf-mimetic superhydrophobic coatings	Micro-/nanostructured waxy surface causing self-cleaning	95% operational efficiency in dusty environments, while significantly reducing the required cleaning frequency	Water repellence, reduced drag	[14]
3	Mussel-inspired adhesive hydrogels	Catechol-based chemistry	Tissue adhesives, marine coatings	Wet adhesion, biocompatibility	[15]

Table 6.1 (continued)

S. no	Bio-inspired material	Mechanism	Application	Performance benefit	References
4	Mosquito-eye inspired coating	Antireflection	Solar panels and low-glare architectural windows, where enhanced light absorption and reduced reflection improve energy performance and visual comfort	12–15% increase in photovoltaic panel efficiency, accompanied by reduced glare and lower optical reflection losses	[16]
5	Beetle-inspired surface coatings	Water collection	Fog-harvesting systems and desert architecture, where enhanced water capture supports passive water collection and climate-resilient building design	Enables up to three times greater water capture compared with smooth surfaces, significantly enhancing moisture collection and surface hydration efficiency	[17]
6	Spider silk-inspired fibers	β-Sheet nanocrystal reinforcement	Biodegradable fibers, tissue scaffolds	Tensile strength ~1 GPa	[18]
7	Gecko-inspired adhesives	Adhesion	Modular building seals and solar installations, where strong, reusable adhesion and improved thermal sealing enhance structural integrity and energy efficiency	Exhibits an adhesion strength of approximately 1 MPa, enables reusable and residue-free bonding, and achieves a 20–30% reduction in heat loss	[19]

6.3 Case Study I: Smart Wearable Energy Systems

6.3.1 Electronic Tongue

Inspired by the structure and functioning of the human tongue, the electronic tongue (e-tongue) is an analytical device designed to recognize and differentiate a wide range of tastes through sensor arrays that emulate biological taste receptors. In the human tongue, taste receptors are unevenly distributed: sour sensations are primarily detected along the sides, bitterness is sensed at the back, and sweetness is perceived at the tip. The fundamental taste modalities—sweet, sour, bitter, salty, and umami—are recognized by specialized receptors [20]. Structurally, vallate papillae are mainly responsible for bitterness perception, fungiform papillae detect most other taste qualities, while filiform papillae provide mechanical support rather than sensory input. In contrast, the e-tongue replicates these biological mechanisms using an array of non-specific but cross-sensitive sensors combined with advanced pattern recognition and chemometric algorithms to generate reliable taste and flavour profiles. Commonly employed sensing platforms include electrochemical, gravimetric, and optical sensors, each contributing complementary information about the chemical composition of the sample [21]. Although direct decoding of taste signals for brain–machine interfaces remain challenging owing to the complexity of interfacing electrodes with taste receptors, gustatory cells, or neural pathways in animals significant progress has been made in chemical sensing applications [22]. Bio-inspired e-tongue systems have demonstrated particular effectiveness in pesticide detection, offering higher selectivity than conventional enzyme inhibition assays. The incorporation of enzymatic elements accelerates targeted recognition reactions, allowing the collection of kinetic information and improving discrimination accuracy during chemometric data analysis. Owing to these advantages, e-tongue platforms integrated with tailored biosensors have found growing use in food quality assessment and agricultural monitoring, where rapid, sensitive, and reliable analysis is essential [23].

Molecularly imprinted polymers (MIPs) represent advanced and reliable recognition elements that are particularly well suited for integration into electronic tongue (e-tongue) systems. Owing to their tailor-made binding sites, MIPs can selectively recognize a wide range of chemical species while maintaining robustness and stability under diverse operating conditions. An important advantage of MIPs is their controlled cross-selectivity, which enhances sensor performance by enabling discrimination among chemically similar analytes a feature that is especially valuable in complex, real-world samples [24]. The versatility of MIP-based sensing platforms supports a broad spectrum of applications, including product development and quality assurance, clinical blood analysis, investigation of molecular interactions, environmental surveillance, and pattern recognition in bioinformatics. When incorporated into e-tongue systems, these materials enable accurate prediction of physiologically and environmentally relevant ions such as Ca^{2+}, Na^{+}, K^{+}, NH_4^{+}, Mg^{2+}, Cl^{-}, SO_4^{2-}, PO_4^{3-}, as well as pH levels. In clinical diagnostics, e-tongues have also demonstrated the ability to quantify metabolites such as urate and creatinine, facilitating routine examinations and the early detection of renal dysfunction through urine analysis. Furthermore, miniaturized pH and K^{+} sensors integrated into e-tongue platforms can be deployed in vivo via endoscopic techniques, allowing real-time monitoring of local biochemical changes. Such approaches have shown promise in identifying ischemic conditions in gastric tissue, highlighting the potential of MIP-enabled e-tongue systems for advanced biomedical diagnostics and personalized healthcare applications.

In addition, e-tongue systems rely on multivariate calibration techniques and advanced pattern recognition algorithms to determine both the qualitative and quantitative composition of complex solutions. This data-driven approach allows e-tongues to extract meaningful chemical information from overlapping and non-specific sensor responses. As a result, these systems have been successfully applied to real-time monitoring of ion and component concentrations in sweat, clinical blood analysis, environmental assessment, evaluation of bitterness in pharmaceutical formulations, and continuous sweat analysis for health tracking. Beyond chemical sensing, electronic tongue technologies have contributed to the development of innovative brain–machine interfaces (BMIs), enabling the creation of brain-controlled prosthetic devices. Such interfaces also provide powerful tools for investigating taste transduction mechanisms and intercellular communication within gustatory pathways. In these experimental frameworks, animals are treated as functional "black boxes," producing measurable outputs in the form of local field potentials and neuronal spike patterns. Using this strategy, fundamental taste stimuli including quinine, sucrose, NaCl, and HCl have been successfully discriminated [25]. Importantly, because analytically useful information can be extracted from decoded signal patterns generated by sensor arrays, the need for highly selective binding sites is reduced. This principle of differential sensing has proven to be an effective strategy for addressing complex analytical challenges, particularly in systems where analytes exhibit similar chemical characteristics or coexist in intricate mixtures [26]. The electronic tongue (e-tongue) has emerged as a valuable tool for monitoring changes in food composition throughout manufacturing processes, particularly during fermentation. Because fermentation is highly sensitive to microbial activity and process conditions, continuous monitoring is essential to detect contamination, prevent process deviations, and ensure consistent product quality. E-tongue systems enable real-time tracking of fermentation dynamics and other compositional changes, providing rapid and objective feedback that supports process control and quality assurance in the food industry. The capability of electronic tongues in food and beverage analysis was clearly demonstrated in 2015 by Blanco et al. [27], who employed a portable e-tongue equipped with electrochemical screen-printed electrodes to analyse lager beers. Using multivariate data analysis and pattern recognition methods, the system successfully predicted key quality attributes such as colour and alcohol content with high accuracy. This work highlighted the practicality of compact e-tongue devices for on-site quality assessment in brewing and beverage production. Similarly, Phat et al. [28] integrated an electronic tongue with conventional chemical analysis and human sensory evaluation to investigate the umami taste of 17 commercially available mushroom varieties. Their study identified aspartic acid, glutamic acid, and 5′-nucleotides as the principal contributors to umami perception, and the equivalent umami concentration (EUC) was subsequently calculated. Strong correlations were observed between EUC values, e-tongue responses, and human sensory scores, demonstrating the reliability of electronic tongue measurements. These findings further suggest that mushrooms represent a promising natural source of umami compounds for use in food additives and industrial flavouring applications.

6.3.2 Electronic Skin

Electronic skin (e-skin) is an advanced sensing platform designed to emulate both the mechanical flexibility and functional capabilities of natural human skin. In addition to mimicking softness, stretchability, and conformability, e-skin systems are capable of detecting a wide range of external stimuli, including humidity, pressure, temperature, and mechanical deformation. To improve the safety, efficiency, and personalized guidance of emerging exercise and rehabilitation technologies, new generations of e-skin sensors have been developed with enhanced sensitivity and adaptability. E-skin has found widespread application in robotics, virtual reality (VR), and healthcare. In medical and assistive technologies, it enables continuous monitoring of physiological parameters and supports the control and feedback of prosthetic devices. In robotics, e-skin enhances environmental interaction by providing tactile feedback, while in VR systems it contributes to more immersive and realistic user experiences. Owing to its ability to form a seamless interface between biological systems and electronic platforms, e-skin is considered a transformative technology for human–machine integration [29]. Current e-skin research incorporates wireless sensing, flexible and stretchable circuits, as well as self-powered and self-healing functionalities to improve durability and autonomy. A wide variety of materials are employed, including hydrogels, conductive polymers, liquid metals, and functional nanomaterials. Numerous sensing devices based on e-skin architectures have been reported for the detection of diverse chemical and biological analytes. In particular, recent advances in iontronics using hydrated and ionic materials such as hydrogels—have shown strong potential for developing self-powered ionic devices. These systems offer multi-stimulus sensing, mechanical robustness, biocompatibility, and the ability to transduce mechanical inputs into electrical signals without the need for external power sources. In parallel, functional oxide materials play a critical role in e-skin and biosensor technologies. Aluminum-doped zinc oxide (ZnO) thin films are widely used as transparent conductive layers in solar cells, large-area flat-panel displays, and liquid crystal displays, and they are especially attractive for electrochemical biosensing applications. Titanium dioxide (TiO_2), a widely used white pigment in paints, inks, plastics, cosmetics, and ceramics, is also highly suitable for biomedical biosensors due to its non-toxicity, biocompatibility, electrical conductivity, and availability in versatile nanostructured forms. Its biocompatible nature supports applications in biosensing devices, light-emitting diodes, and ultraviolet photodetectors, while its wide band-gap semiconductor properties enable functionality across a broad range of technological applications [30]. According to the work of El-Shaer et al. [31] potassium-doped p-type Cu_2O thin films exhibit enhanced physical and photoelectrochemical properties, making them highly suitable for biosensor applications. The introduction of K dopants improves electrical conductivity and promotes more efficient separation of photogenerated charge carriers, thereby increasing sensor sensitivity and performance under photoelectrochemical operation [31]. In biomedical e-skin systems, force and pressure sensors play a critical role in improving prosthetic functionality by enabling precise force feedback, while pressure-mapping technologies help identify high-risk pressure points in immobile patients, reducing the likelihood of bedsores. Integrated temperature sensors further support healthcare monitoring by allowing early detection of inflammation and infection. In parallel, visual display components embedded within wearable technologies provide users with immediate access to physiological data and alerts directly on the skin. Owing to their high sensitivity and spatial resolution, e-skin-based biomedical sensors enable accurate data acquisition, supporting effective and personalized therapeutic interventions. The incorporation of self-healing capabilities significantly enhances the durability, reliability, and long-term usability of e-skin devices. Moreover, the integration of wireless communication technologies improves usability and convenience, allowing seamless data transmission to external devices. In wound care, remote monitoring systems coupled with smartphone applications enable patients and caregivers to perform routine wound assessments at home. This facilitates early intervention, timely adjustment of treatment strategies, and improved patient satisfaction. Collectively, these advances in electronic skin technologies hold substantial promise for improving wound management, enabling early detection of complications, and supporting personalized wound-healing therapies [29]. Several pioneering developments further illustrate the potential of e-skin technologies. The University of Toronto has developed an artificial ionic skin capable of reproducing complex tactile sensations of human skin, significantly enhancing prosthetic performance and human–machine interaction. Binghamton University introduced a wound-monitoring sensor that delivers real-time feedback on wound conditions, aiding optimal clinical decision-making. At Yonsei University, an interactive skin display capable of monitoring sweat gland activity and skin temperature has been developed for assessing stress and physical exertion. Stanford University's Body Net sensor measures respiration and pulse rates, making it well suited for continuous and remote health monitoring applications [29]. Additional innovations include a wireless wearable device from the Georgia Institute of Technology that enables comprehensive monitoring of electrocardiograms (ECG), heart rate, and respiratory rate for both clinical and personal healthcare use. Similarly, an electronic tattoo (e-tattoo) developed by the University of Texas at Austin integrates seismocardiography and ECG signals to provide a detailed evaluation of cardiac health. Together, these advanced sensing platforms represent significant milestones in wearable healthcare technology and have the potential to fundamentally transform medical diagnostics and monitoring. Looking ahead, emerging research suggests that healthcare-oriented e-skin sensors will play a transformative role in future therapy and patient monitoring. These systems are expected to enable personalized pain management, advanced wound-care support, and real-time monitoring of vital signs. Furthermore, e-skin platforms will facilitate remote patient monitoring, integration with controlled drug-delivery systems, enhancement of prosthetic devices, and early disease diagnosis through biomarker detection. Collectively, these developments are poised to improve patient outcomes and drive more efficient, responsive, and personalized medical treatments [32].

6.4 Case Study II: Off Grid Rural Power Units

The natural world's solutions have served as a source of inspiration for researchers' ideas and breakthroughs. In many ways, including economic growth, energy plays a critical role in the sustainable evolution of humanity. Researchers can be motivated by nature to create innovative technology for contemporary sustainable energy solutions as shown in Fig. 6.2 [33].

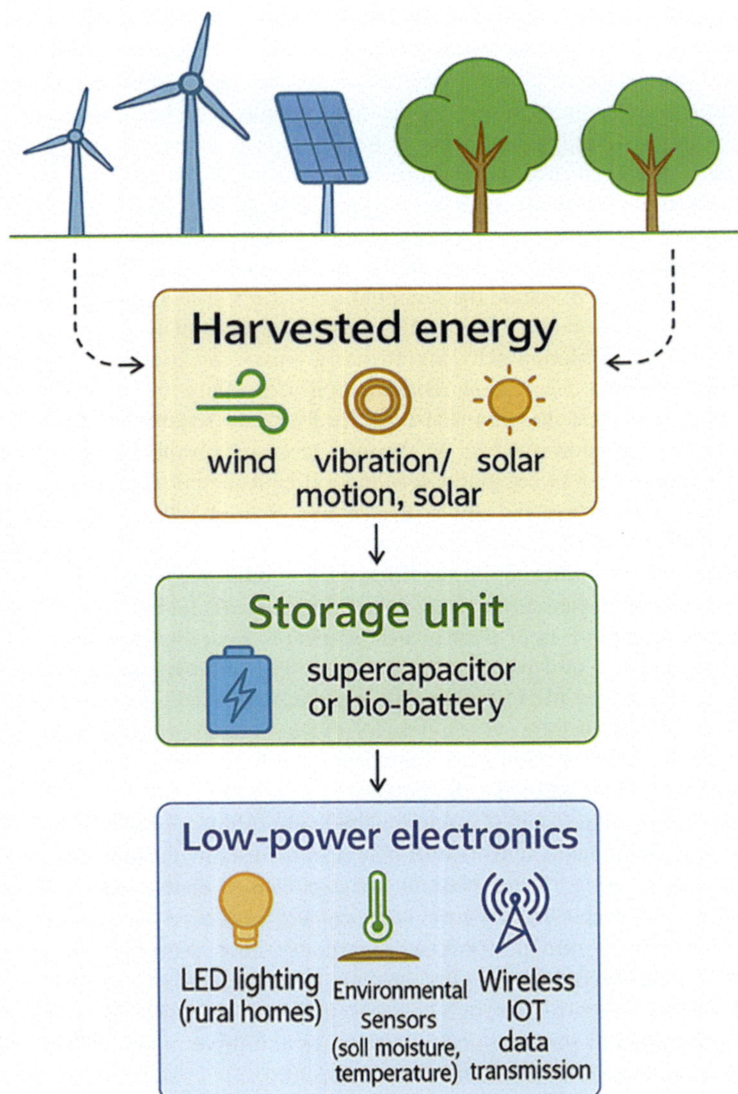

Fig. 6.2 Functional layout of bio-inspired off-grid power modules

Bio-inspired technological innovations are increasingly contributing to the fulfillment of global energy demands, particularly in the field of solar energy, through devices such as dye-sensitized solar cells (DSSCs) and bio-inspired antireflective (AR) structures integrated into solar cell electrodes [34]. Drawing inspiration from natural grating-like structures found on the exoskeleton of the Oriental hornet and in the compound eyes of moths, researchers have incorporated a hierarchical AR layer into conventional TiO_2 electrodes of DSSCs. This design enhances light harvesting by combining AR effects with diffraction, enabling absorption over a broader range of wavelengths. As a result, the power conversion efficiency (PCE) of DSSCs equipped with a 2D patterned TiO_2 AR layer increased from 5% to 6.5%. The bio-inspired AR layer is created by integrating a hierarchically structured diffraction grating onto the standard TiO_2 working electrode. To optimize photon-trapping effects in the visible spectrum, TiO_2 nanowires (NWs) were fabricated using a polymer template shaped with circular holes via interference lithography (IL). Subsequently, a layer of TiO_2 nanocrystals was deposited on the nanowires using the doctor-blade technique. Platinum was used as the counter electrode in these DSSC assemblies, completing the cell architecture [35]. More broadly, the study of bio-inspired dyes, organisms, and structural motifs is guiding the design of highly efficient light-harvesting devices, including solar cells, solar concentrators, and solar power satellites. Certain biological systems naturally exhibit remarkable optical or thermal properties that can be emulated to enhance energy capture and conversion. By mimicking these structures—both living and nonliving—researchers aim to develop more stable, efficient, and robust solar energy technologies. Solar cells that incorporate such biologically inspired strategies are collectively referred to as bio-inspired solar cells (BISCs) [36].

In order to create next-generation PV technologies, BISCs are manufactured by imitating the structures and arrangements of biological systems while simultaneously using the latest developments in nanotechnology. The development of several scientific technologies opens the door to the creation of artificial functions that are as exact as those found in organic things. In order to create BISCs, biological structures and functions must be artificially created. The design and development of reliable and effective solar PV systems is driven by photosynthesis and the cyanobacteria energy conversion process from sunlight to produce food in plants and insects [37]. Researchers developed bio-inspired light trapping mechanisms for electrical energy conversion using photovoltaic cells based on the energy recovery process of photosynthesis [37]. In order to create photovoltaic cells, efforts are being made to develop artificial biological structures and functionalities using various building materials. Certain insects have compound eyes made up of densely packed papillae in the shape of a hexagon, known as ommatidia. Internal photoreceptors in ommatidia are encircled by pigment and support cells. These components efficiently reduce optical reflectance in insects' eyes [38]. Scientists were inspired by the insect's compound sight to create high absorption and antireflective solar cells that use nanotechnology and other technologies to limit sunlight reflection to 35%. The full explanation of the antireflective phenomena would come from a thorough examination of the structures on the surface of moth eyes [39]. Moths' capacity to see at night would be enhanced by the highly ordered AR nanostructured array of 200–300 nm-sized pillars on their surface, which would reduce light reflection in low light. Antireflective (AR) nanostructures are typically smaller than visible light wavelengths (380–760 nm), preventing them from causing an abrupt refractive index change at the surface. To overcome this, subwavelength structures are designed to produce a gradual refractive index transition between air and the underlying medium, a principle that underlies the moth-eye effect. Biomimetic moth-eye structures, often featuring paraboloid profiles and a continuously varying effective refractive index (ηeff), progressively alter the refractive index of incident light, thereby minimizing reflection [40]. Such designs have been widely applied to improve optical performance in devices like solar cells. A simple bio-inspired fabrication method, polymerization-induced self-wrinkling, has been used to create AR coatings with micro-wrinkle patterns. During photo-crosslinking, wrinkle microstructures spontaneously form on the coating's surface, facilitating enhanced light trapping and photon absorption. In these coatings, acrylate cross-linked trimethylpropane triacrylate, with 0.3 wt% photoinitiator, was blended with fluorinated copolymers composed of monomers such as methyl methacrylate (PMMA-F), n-butylacrylate (PBA-F), styrene (PS-F), and poly (ethylene oxide) dimethacrylate (PEGMA-F). The solution was applied to α-Si solar cells using a knife-coating technique. Encapsulation with wrinkle-containing AR coatings enhanced photovoltaic performance by 4–8%, with transmittance reaching 90%, reflection reduced by 5–8%, and AR efficiency improved by 4–7%. The wrinkle patterns increase the optical path length in the photoactive layer and reduce light reflection, effectively boosting light-harvesting efficiency. Beyond structural coatings, bio-inspired strategies have guided the development of transparent conducting electrodes (TCEs) for organic solar cells (OSCs). Conventional TCEs are often limited by poor mechanical and chemical stability. Graphene, with its superior optical, electrical, and physical properties and inherent hydrophobicity, was modified with a thin poly(norepinephrine) layer—an amphiphilic catecholamine derivative—to make the surface hydrophilic without sacrificing transmittance or conductivity. OSCs using poly(norepinephrine)-coated graphene electrodes achieved a PCE of 7.93%, comparable to 8.73% for ITO-based reference devices, demonstrating an effective bio-inspired surface modification for chemically inert graphene. Additional bio-inspired approaches for OSCs include the design of boron difluoride complexes of curcuminoid derivatives end-capped with triphenylamine groups. These materials improved PCE, chemical stability, and thermal robustness [41]. In perovskite solar cells, biological materials such as bacteriorhodopsin (bR) proteins have been used to enhance performance. By aligning the energy gaps between bR proteins and perovskite materials via Förster resonance energy transfer (FRET), the PCE increased from 14.59% to 17.02% compared to conventional cells [42]. Similarly, copolymers derived from isindigo and chlorophyll derivatives have been shown to achieve high PCEs in OSCs. Further improvements in light harvesting have been achieved by integrating natural photosynthetic complexes at the interface between active layers and charge-collecting electrodes. For example, silver nano prisms combined with photosynthetic materials have enabled OSCs with PCEs exceeding

10% [43]. Collectively, these bio-inspired strategies from subwavelength AR nanostructures to functionalized electrodes and natural light-harvesting molecules demonstrate the powerful role of biomimicry in enhancing solar cell efficiency, stability, and overall performance.

6.5 Conclusion and Future Perspective

A revolutionary step towards sustainable energy and sensing solutions is the combination of self-powered system design with bio-inspired nanomaterials. These systems offer effective energy harvesting, storage, and utilization in both smart wearables and decentralized off-grid applications by imitating natural structures and features including hierarchical structuring, adaptive response, and energy conversion. Chitin-, cellulose-, and protein-based nanostructures have shown great promise for flexible, biocompatible, and environmentally friendly power production when integrated into triboelectric, piezoelectric, and photovoltaic systems. Large-scale production, long-term stability, and smooth system integration with cutting-edge electronics and communication networks continue to present difficulties, nevertheless. To create scalable, eco-friendly energy platforms, future research should focus on circular bioeconomy models, AI-assisted materials optimization, and multifunctional hybrid nanostructures. In the end, the advancement of self-sufficient technologies for wearable electronics of the future and sustainable rural energy ecosystems will depend heavily on the collaboration between bio-inspired material innovation and system level engineering.

References

1. Xin W, Jiang L, Wen L (2022) Engineering bio-inspired self-assembled nanochannels for smart ion transport. Angew Chem Int Ed 61(40):e202207369
2. Johnson AP, Sabu C, Nivitha KP, Sankar R, Ameena Shirin VK, Henna TK, Raphey VR, Gangadharappa HV, Kotta S, Pramod K (2022) Bio-inspired and biomimetic micro- and nanostructures in biomedicine. J Control Release 343:724–754. https://doi.org/10.1016/j.jconrel.2022.02.013
3. Chang BP, You X, Andrzejewski J, Skorczewska K (2025) Editorial: renewable biosourced carbon materials derived from biomass and their biocomposites fabrication for innovative applications. Front Mater 11. https://doi.org/10.3389/fmats.202
4. Shreya R (2024) Implementation of biomimicry for advanced impact-resistant composites: advanced manufacturing techniques, pp 166–179. https://doi.org/10.1201/9781
5. Moussaei M, Haddadi-Asl V, Ahmadi H (2024) Ternary nanocomposite of RAFT polymerized DMAEMA-functionalized graphene oxide, manganese dioxide, and polyaniline as active electrode material for supercapacitor. Polym Int 73:612–624. https://doi.org/10.1002/pi.6631

6. Zheng Y, Wang J, Wang J, Li Y, Jiang Z (2025) Insect cuticle: a source of inspiration for biomimetic Interface material design. Colloid Interface Sci Commun 64:100818. https://doi.org/10.1016/j.colcom.2025.100818
7. Zhao F, Zhan F, Wang L (2022) Hybrid topography of lotus leaf under hydrostatic/hydrodynamic pressure. Adv Mater Interfaces 10(4):2202044. https://doi.org/10.1002/admi.2022020
8. Subramani C, Usha S, Jayakumar S, Palanisamy R, Sarawagi R, Singh AK (2024) Solar panel self-cleaning mechanisms and their effect on economic and environmental sustainability. J Electr Comput Eng 2024(1). https://doi.org/10.1155/2024/772671
9. Yu J (2024) Pathways in biomimicry to enhance solar technology capabilities. Adv Econ Manag Polit Sci 104(1):134–142. https://doi.org/10.54254/2754-1169/104/2024mur0126
10. Jessop AL, Pirih P, Wang L, Patel NH, Clode PL, Schröder-Turk GE, Wilts BD (2024) Elucidating nanostructural organization and photonic properties of butterfly wing scales using hyperspectral microscopy. J R Soc Interface 21(218):20240185. https://doi.org/10.1098/rsif.2024.0185
11. Omidian H (2024) AI-powered breakthroughs in material science and biomedical polymers. J Bioact Compat Polym 40(10). https://doi.org/10.1177/0883911
12. Qu L, Gou Q-Q, Deng J, Zheng YJ, Li M (2024) A perspective of bio-inspired interfaces applied in renewable energy storage and conversion devices. Langmuir 40(10). https://doi.org/10.1021/acs.langmuir.3c03679
13. Zhang Q, Song K, Liu J, Dong J, Liang Y, Du T, Fan H (2025) Tailorable nanoconfinement enables nacreous biomimetic graphene/silicate composites with ultrahigh strength and toughness. https://doi.org/10.2139/ssrn.5085730
14. Dhoska K, Spaho E, Sinani U (2024) Fabrication of black silicon antireflection coatings to enhance light harvesting in photovoltaics. Eng 5(4):3358–3380. https://doi.org/10.3390/eng5040175
15. Xu S, Kang M, Xin X, Liang J, Xiao H, Lu Y, Yang J, Zhai H (2024) Design principles and application research of mussel-inspired materials: a review. J Environ Chem Eng 12(1):111655
16. Alqarni AS, Bahattab MA, Alqahtani MHM, Almater A, Abaalkheel I, Aldhuwaile AAA (2024) Development of anti-dust nanostructured silicon dioxide coating for solar cells. https://doi.org/10.20944/preprints202412.0469.v1
17. Tu S, Zhang L, Tu S, Feng J (2025) Waterborne recoatable transparent superhydrophobic coatings with excellent self-cleaning and anti-dust performance. Small. https://doi.org/10.1002/smll.202410171
18. Vukusic P, Sambles J (2003) Photonic structures in biology. Nature 424:852–855. https://doi.org/10.1038/nature01941
19. Fina J, Kaur N, Chang C-Y, Lai CS, Radu DR (2023) Enhancing light harvesting in dye-sensitized solar cells through mesoporous silica nanoparticle-mediated diffuse scattering back reflectors. Electron Mater 4(3):124–135. https://doi.org/10.3390/electronicmat40
20. Vahdatiyekta P, Zniber M, Bobacka J, Huynh TP (2022) A review on conjugated polymer-based electronic tongues. Anal Chim Acta 1221:340114. https://doi.org/10.1016/j.aca.2022.340114
21. del Valle M (2012) Sensor arrays and electronic tongue systems. Int J Electrochem 2012:986025. https://doi.org/10.1155/2012/986025
22. Hsia KJ, Wu C, Liu Q (2015) Bio-inspired smell and taste sensors. https://doi.org/10.1007/978-94-017-7333-1
23. Khan MRR, Khalilian A, Kang SW (2016) A high sensitivity IDC-electronic tongue using dielectric/sensing membranes with solvatochromic dyes. Sensors 16(5). https://doi.org/10.3390/s16050668
24. Podrażka M, Bączyńska E, Kundys M, Jeleń PS, Witkowska Nery E (2018) Electronic tongue—a tool for all tastes? Biosensors 8(1):3. https://doi.org/10.3390/bios8010003

25. Qin Z, Zhang B, Hu L, Zhuang L, Hu N, Wang P (2016) A novel bioelectronic tongue in vivo for highly sensitive bitterness detection with brain-machine interface. Biosens Bioelectron 78. https://doi.org/10.1016/j.bios.2015.11.078
26. Kalinowska A, Wicik M, Matusiak P, Ciosek-Skibinska P (2022) Chemosensory optode array based on pluronic-stabilized microspheres for differential sensing. Chem 10(1). https://doi.org/10.3390/chemosensors10010002
27. Blanco CA, De La Fuente R, Caballero I, Rodríguez-M'endez ML (2015) Beer discrimination using a portable electronic tongue based on screen-printed electrodes. J Food Eng 157(1). https://doi.org/10.1016/j.jfoodeng.2015.02.018
28. Phat C, Moon B, Lee C (2016) Evaluation of umami taste in mushroom extracts by chemical analysis, sensory evaluation, and an electronic tongue system. Food Chem 192:1068. https://doi.org/10.1016/j.foodchem.2015.07.113
29. Youn S, Ki MR, Abdelhamid MAA, Pack SP (2024) Biomimetic materials for skin tissue regeneration and electronic skin. Biomimetics. https://doi.org/10.3390/biomimetics9050278
30. Ismail W, Ibrahim G, Atta H, Sun B, El-Shaer A, Abdelfatah M (2024) Improvement physical and photoelectrochemical properties of TiO_2 nanorods toward biosensor and optoelectronic applications. Ceram Int 50(10):17968–17976. https://doi.org/10.1016/j.ceramint.2024.02.286
31. El-Shaer A, Darwesh N, Habib MA, Abdelfatah M (2024) Doping of nanostructured Cu2O films to improve physical and photoelectrochemical properties as a step forward for optoelectronics applications. Opt. Mater. (Amst) 148:114849. https://doi.org/10.1016/j.optmat.2024.114849
32. Pereira AN, Noushin T, Tabassum S (2021) A wearable, multiplexed sensor for realtime and in-situ monitoring of wound biomarkers. In: Proceedings of IEEE Sensors. https://doi.org/10.1109/SENSORS47087.2021.9639722
33. Ravi SK, Tan SC (2015) Progress and perspectives in exploiting photosynthetic biomolecules for solar energy harnessing. Energy Environ Sci 8:2551–2573
34. Huang Z, Cai C, Kuai L, Li T, Huttula M, Cao W (2018) Leaf-structure patterning for antireflective and self-cleaning surfaces on Si based solar cells. Sol Energy 159:733–741
35. Ahn HJ, Kim SI, Yoon JC, Lee JS, Jang JH (2012) Power conversion efficiency enhancement based on the bio-inspired hierarchical antireflection layer in dye sensitized solar cells. Nanoscale 4:4464
36. Zhang Y, Mei J, Yan C, Liao T, Bell J, Sun Z (2019) Bio-inspired 2D nanomaterials for sustainable applications. Adv Mater 32:1902806
37. Makita H, Hastings G (2017) Inverted-region electron transfer as a mechanism for enhancing photosynthetic solar energy conversion efficiency. PNAS 114(35):9267–9272
38. Kuo WK, Hsu JJ, Nien CK, Yu HH (2016) Moth-eye-inspired biophotonic surfaces with antireflective and hydrophobic characteristics. ACS Appl Mater Interfaces 8:32021–32030
39. Dong C, Yu HK, Shen KS, Zhang J, Xia SQ, Xiong ZG et al (2018) Low emissivity double sides antireflection coatings for silicon wafer at infrared region. J Alloys Compd 742:729_735
40. Sun CH, Jiang PH, Jiang B (2008) Broadband moth-eye antireflection coatings on silicon. Appl Phys Lett 92(6):061112
41. Archet F, Yao D, Chambon S, Abbas M, D'Aleo A, Canard G et al (2017) Synthesis of bio-inspired curcuminoid small molecules for solution-processed organic solar cells with high open-circuit voltage. ACS Energy Lett 2(6):1303–1307
42. Das S, Wu C, Song Z, Hou Y, Koch R, Somasundaran P et al (2019) Bacteriorhodopsin enhances efficiency of perovskite solar cells. ACS Appl Mater Interfaces 11:30728–30734
43. Vohra V (2018) Natural dyes and their derivatives integrated into organic solar cells. Materials 11:2579

Chapter 7
Sustainability Assessment of Bio-Inspired Energy Materials: Life Cycle and Environmental Perspectives

Abstract New avenues for sustainable energy production, storage, and conversion have been made possible by the quick development of bio-inspired energy materials. These materials provide creative ways to lessen reliance on non-renewable resources and lessen environmental effects since they are generated from or imitate natural systems. Using thorough life cycle assessment (LCA) and environmental evaluation methodologies, this chapter investigates the sustainability of bio-inspired energy materials. It emphasizes how crucial it is to take into account every stage of the material life cycle, from the extraction and synthesis of raw materials to application, end-of-life treatment and recycling. To demonstrate practical uses and difficulties in incorporating bio-inspired materials into energy systems, case examples are also provided.

Keywords Bio-inspired materials · Synergistic approaches · Life cycle assessment · Environmental impact assessment · Sustainability

7.1 Introduction

Nature is a master craftsman that has been perfecting structures for a very long time. In order for living things to adapt to their hostile surroundings, evolution has produced a variety of structures with distinct roles and characteristics. Natural selection and ongoing evolution have improved the functionality of living creatures' surfaces. Scientists have developed new artificial surfaces and materials that have the same or better properties than their counterparts because of the mechanism underlying the intriguing properties of valuable biological materials. The word "bio-inspiration" refers to attempts to recognize, create, and imitate natural things in order to get a deeper comprehension of nature [1]. The term "inspiration" describes the initial stage of design or function observation that yields concepts for creating comparable products. With the primary goal of achieving sustainability, mimicry is an advanced type of inspiration that uses a variety of technology instruments to create materials that resemble natural items. According to Zhang and Le Ferrand [2], bio-inspiration

M. Y. Mir, J. A. Parray, *Nanoenergy Production*, SpringerBriefs in Energy,
https://doi.org/10.1007/978-3-032-20859-0_7

holds great promise for advancing sustainable development. Biomimetics and bio-inspiration are rapidly developing applied fields where biological scientific knowledge can be used to solve design problems. Otto Herbert Schmitt used the word "biomimetics"in 1957 to refer to the transfer of information from biological systems to engineering or design. Over the past 20 years, bio-inspiration and biomimetics research has increased. The term "biomimetics"has evolved into "bio-inspiration," which refers to the process of creating new ideas inspired by natural items. Although the two concepts are similar, they take different approaches, and it has been noted that knowledge transmission is not necessary. For instance, soft robotics can mimic a butterfly's wing flapping or proboscis extension [3]. Nonetheless, bio-informed methods demonstrate a comprehension of the intricate mechanisms or processes of biological systems, which inspires creative design. The nose of a bullet train is shaped like a bullet, just like the beak of a kingfisher. Because the beak travels quickly from the air to the water, animals can use it to capture aquatic prey.

Although the timeframes for engineering and natural things differ, their goals such as cost-effectiveness, multifunctionality, optimization, and design constraints are similar [4]. While engineered materials rely on their qualities to build design, natural materials are superior because of their structural hierarchical integrative material system. The formation of natural materials is a lengthy, bottom-up process in which materials evolve, self-assemble, and adapt to their surroundings. Because bottom-up techniques are based on nature-triggered protocol, they are durable and allow natural things to be hierarchical at all scales. The hierarchical structure of biological materials, such wood, bone, and teeth, allows for greater strength with less material [5]. The manufacture of correctly ordered biomaterials, such nacre and bone, which are controlled by living systems through large-scale manufacturing techniques, is challenging to replicate [4]. By applying particular nanostructure properties and biological molecules' functions, bio-inspired synthesis provides an alternative environmentally friendly method for producing functional bio-nanomaterials. Biomolecular identification and aggregation processes help define the role and shape of bio-fabricated substances [6]. The bio-inspired or biomimetic syntheses have several advantages over traditional chemical synthesis methods. Compared to conventional material processing techniques, the size and shape of the required bio-inspired nanoparticles may be easily controlled, and the reaction conditions are moderate at room temperature. Biomolecules have important roles in the alteration of structure and the function of bio-fabricated devices, as well as unique aggregation and molecular identification characteristics. Researchers are creating multifunctional mimetics, such as self-cleaning buildings whose sections are strengthened by termite mounds or crystal structures [7].

7.2 Life Cycle Assessment of Bio-Inspired Energy Materials

The development of a comprehensive sustainability assessment model to support the targeted design of sustainable products is founded on a set of methodological and application-oriented requirements. At the most general level, these requirements reflect the core principles of scientific methodology, namely consistency, comparability, reproducibility, and falsifiability. Beyond these generic criteria, which apply to all assessment frameworks, additional context-specific requirements must be considered. These relate in particular to bio-inspired systems, sustainability assessment methodologies, and decision support systems (DSS) [8]. Although decision theory primarily focuses on the organization and management of decision-making processes, it offers several broadly applicable requirements for DSS design. A common theme across these approaches is the emphasis on adaptability and flexible applicability. Accordingly, a DSS should be capable of supporting both semi-structured and unstructured decisions, be usable by decision makers with varying levels of expertise, and provide support at any stage of the decision-making process [9]. Finkbeiner et al. [10] proposed a semi-quantitative evaluation framework consisting of seven criteria that capture the most relevant aspects commonly addressed in scientific literature. Each criterion is subdivided into three performance levels, resulting in a structured scorecard for assessing sustainability evaluation systems. From a bio-inspired perspective, an assessment system can be characterized by the intrinsic properties of effectiveness, flexibility, and resilience [11]. Effectiveness reflects the effort required from practitioners to generate a given amount of meaningful information. As this aspect can only be reliably assessed through practical application, a sufficient number of applied studies integrating Bio-inspired Sustainability Assessment (BiSA) are required. Flexibility is achieved when the system allows for adaptable and expandable indicator sets and weighting schemes, supported by continuous self-evaluation and iterative refinement. Finally, the assessment model is considered resilient if it is able to accommodate changes in input data by responding with appropriate and proportionate adjustments, rather than failing or producing unstable results [12].

First and foremost, sustainability can be understood as a societal paradigm that has the potential to influence almost all levels of decision-making, ranging from everyday individual choices to international policy and governance. Owing to its constructive ambiguity, high level of abstraction, complex cause–effect relationships, strong interconnectedness, and multidimensional nature, the meaning of sustainability is often shaped and reinterpreted through subjective perspectives [13]. As a result, the concept remains fluid and context dependent. Although sustainability has gained widespread acceptance and prominence, there is still a lack of agreement on specific concepts that extend beyond the broad consensus established in *Our Common Future*. While the Sustainable Development Goals (SDGs) mark an important step toward global alignment, they do not yet offer a complete, operational, and quantifiable set of indicators suitable for systematically assessing sustainability, particularly in the context of the intentional design and development of sustainable products. Overall, the diversity of definitions, interpretations, and associated assessment approaches indicates that the sustainability paradigm is still evolving, and its definitive structure and methodological foundations have yet to be fully established [14].

The following discussion reviews key concepts and assessment approaches with a particular focus on their suitability and internal consistency for the quantitative evaluation of products over their entire life cycle. Although a wide range of sustainability concepts exists, most remain largely conceptual or schematic in nature. To enable rigorous life cycle thinking and generate results that are comparable, transparent, and sufficiently detailed for informed decision-making, these concepts must be translated into differentiated and quantifiable assessment frameworks [15]. For example, the "cradle-to-cradle" concept is frequently proposed as a guiding design philosophy for sustainable products; however, when integrated with quantitative life cycle assessment methods, it reveals several conceptual and methodological inconsistencies [16]. A similar limitation applies to the concept of natural capitalism, which, in practice, functions more as a single-score valuation approach within Life Cycle Assessment (LCA) than as a comprehensive sustainability framework, as it primarily monetizes environmental resources. While the application of human and manufactured capital within triple-bottom-line LCA allows all three sustainability pillars to be considered, their evaluation remains largely confined to monetary metrics, restricting interpretative depth [17]. Many existing sustainability assessment methods are neither quantitatively comparable nor methodologically compatible with one another. These approaches are therefore reviewed primarily to establish the current state of research, as several recent studies have already provided comprehensive overviews of commonly applied sustainability assessment frameworks [18]. In particular, Guinée [18] presents an extensive meta-evaluation of Life Cycle Sustainability Assessment (LCSA) studies reported in the scientific literature. This analysis identifies numerous general recommendations and highlights methodological gaps that persist in current practice and should be addressed in the further development of LCSA. Key issues raised include the availability and quality of data and methods—especially for social indicators—the effective communication of results, the integration of positive contributions alongside burdens, and the avoidance of double counting and inconsistent methodological application. As many of these challenges remain unresolved, there is strong justification for continued methodological advancement aimed at improving sustainability assessments in terms of robustness, transparency, and communicative power, as well as expanding their analytical breadth and depth [18]. A detailed overview of the life cycle sustainability performance of emerging bio-inspired energy materials is presented in Table 7.1.

Table 7.1 Sustainability and life-cycle assessment of bio-inspired energy materials

Material	Key environmental concerns	LCA observations	Sustainability strategies	References
Triboelectric Nanogenerator (TENGs) — Polymer or biopolymer based, bio-inspired structures	Embodied energy and emissions from polymer production & fabrication; plastic waste; solvent/ chemicals; end-of-life (EoL) disposal	Use of recycled polymers or biopolymers shifts LCA favourably; recycled-PET based TENGs reduce carbon footprint vs. virgin-PET	Use recycled plastics (e.g. PET/ PP), biopolymers; design for recyclability; adopt circular-economy EoL loops	[19]
Piezoelectric energy harvester (ceramic or composite piezo materials / thin-films)	High energy demand during ceramic synthesis (e.g. sintering); use of heavy/toxic elements (e.g. lead in PZT); electrode deposition; difficult recycling of ceramics	For traditional lead-based piezo (e.g. PZT), LCA shows high environmental & toxicity impacts over life cycle; lead-free alternatives (e.g. perovskite/polymer composites) may significantly reduce impact	Favor lead-free piezo materials; low-temperature processing; polymer/composite-based harvesters; avoid heavy metal ceramics	[20]
Bio-based / biopolymer-derived hydrogels / materials (e.g. cellulose-based) for sensors / substrates / energy-related components	Environmental load from biomass sourcing; water/ solvent use; chemical treatments; variability in biodegradation, EoL fate	Biopolymer-based systems (e.g. cellulose) can lower fossil-depletion and embodied GHG compared to petro-polymers; when using green-chemistry approaches, overall footprint is reduced	Use sustainably sourced biomass; minimize solvent/ chemical use; prioritize water-based / green-chemistry processing; design for biodegradability or safe composting	[21]
Biomimetic conductive polymers / composites / bio-templated electrodes (e.g. polymers, MXene- or biotemplate-based) for energy devices	Chemical toxicity and embodied energy of monomer synthesis, dopants, solvents; energy for exfoliation or processing (for e.g. MXenes); limited EoL / recycling data	When using water-based polymerization and green dopants/ solvents, environmental footprint reduces substantially compared to conventional routes — But comprehensive, peer-reviewed LCA for many lab-scale biomimetic materials still scarce	Prioritize aqueous processing, low-toxicity chemicals; design for disassembly and recyclability; develop EoL/ recycling protocols; build life-cycle inventories for biomimetic lab materials	[22]

(continued)

Table 7.1 (continued)

Material	Key environmental concerns	LCA observations	Sustainability strategies	References
Recycled-plastic / circular-economy TENGs / devices (e.g. recycled PET/PP, waste-stream plastics)	Quality degradation of recycled plastics; energy & emissions of recycling processes; variability/ fluctuation in waste supply; end-of-life microplastics / further waste	Using recycled PET for TENGs manufacturing reduces life-cycle carbon footprint compared to virgin PET; circular-loops offer improved material sustainability if recycling and waste-management infrastructures are effective.	Design devices for mono-material construction; ensure recyclability; adopt closed-loop collection/re-use; assess recyclate performance over multiple cycles; integrate circular-economy perspective	[23]

7.3 Environmental Impact Assessment of Bio-Inspired Energy Materials

A central question in bio-inspired research is whether living nature can truly be regarded as a complete and direct analogue of what is commonly defined as sustainability. This issue is critical when attempting to derive sustainable solutions from biological systems. The underlying premise is that, if nature is adopted as a direct model, the strategies observed in natural systems should have evolved under constraints comparable to those used in contemporary sustainability assessments. If this condition is not met, then the direct transfer of biological principles cannot automatically be considered sustainable. In such a case, sustainability must be recognized as a largely human-constructed concept that is, at least to some extent, distinct from our scientific understanding of natural systems. Addressing this question therefore requires a careful examination of both the conceptual foundations of sustainability and the core operating principles of biological systems. When efforts are made to explicitly align biological systems with sustainability indicators, two major inconsistencies become apparent. First, the prevailing definition of sustainability is inherently anthropocentric and is not rooted in biological logic. This is evident in many sustainability frameworks, including those promoted by the United Nations, which explicitly prioritize human well-being and societal goals and employ sustainability primarily as an evaluative tool for human activities [24]. The second inconsistency arises from the notion of social sustainability, which remains conceptually unresolved and contested within the sustainability discourse [25]. Given these limitations, sustainability viewed through a bio-inspired lens can be interpreted as the interdependence of system functions combined with an unavoidable consumption of resources. From this perspective, a system may be considered sustainable if it fulfils its required functions while simultaneously ensuring the long-term and timely availability of the resources it depends on. This interpretation highlights a dominant

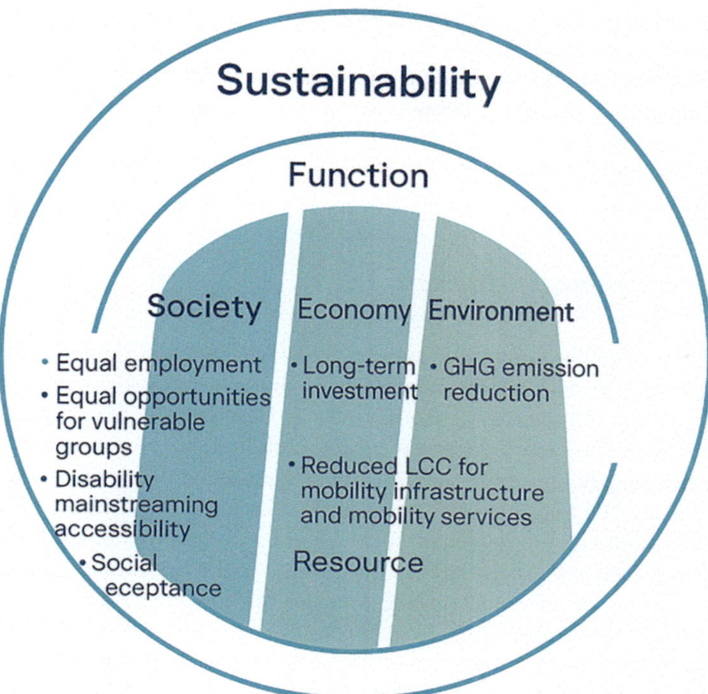

Fig. 7.1 Conceptual framework of bio-inspired sustainability assessment, integrating the three dimensions society, economy, and environment with function and resource aspects that link societal functions to underlying environmental and economic resources

trend in modern economic metabolism: the organization of social systems through the extraction and consumption of environmental resources, driven by economic infrastructures and processes (Fig. 7.1). The graph is designed to illustrate the dynamics of this metabolic system, indicating that sustainability is achieved when the overall form of the system is preserved, thereby ensuring a continuous availability of resources to support the ongoing generation of functions. Within this framework, the economy is viewed as a means rather than an end, transforming resources into functions that enable business models and facilitate the adoption of new goods and services. Because many processes are driven by motivations that extend beyond this dominant flow direction, it is not possible to define a single, universal rule governing all interactions. Instead, the concept functions as a conceptual model that helps explain how activities accomplish both their intended outcomes and their unintended consequences. In practice, processes are rarely designed to deliberately generate environmental functions or to use environmental resources in a non-depletive manner. The economy's primary role is therefore represented as an interface linking society and the environment. To sustain this metabolism, it is crucial to prevent resource exploitation from exceeding levels that would undermine the continued performance of societal functions. Under this conceptual scheme, a system can be regarded as sustainable if the dynamic societal metabolism is maintained and remains stable despite changing conditions. The model offers a shell-like structure capable of accommodating a range of assessment approaches, while also extending existing frameworks by explicitly incorporating positive contributions alongside conventional evaluative elements.

7.3.1 Environmental Burden

Because the depletion of natural resources underpins much of human prosperity, it must be carefully accounted for when assessing bio-inspired sustainability. Drawing inspiration from biological systems, the most evident parallel lies in their reliance on physical resource inputs. As noted earlier, both biotic (living) and abiotic (non-living) components of the environment are indispensable to ecosystem functioning. Throughout Earth's history, source–sink dynamics have been shown to influence variations in habitat quality, which in turn affect biodiversity, population dynamics, and species abundance, particularly in the aftermath of mass extinction events [26]. While it is often claimed that nature operates without generating waste, closer examination suggests otherwise. For example, a substantial portion of today's crude oil originates from marine organisms that died and were buried under sedimentary layers over geological timescales. These deposits can therefore be interpreted as long-term repositories of biological waste, or fossilized organic matter. Consequently, the extraction and use of fossil resources such as coal, natural gas, and crude oil hydrates are inherently associated with carbon dioxide emissions. This perspective challenges the simplistic notion of "zero waste" in nature and highlights the importance of temporal and spatial scales when interpreting natural cycles. Life cycle assessment (LCA) typically quantifies environmental impacts using a single-point indicator derived from multiple characterization methods. However, due to the interactive and system-oriented nature of Bio-inspired Sustainability Assessment (BiSA), it is not feasible to fully apply standardized LCA procedures as prescribed in existing norms [26]. Despite this limitation, several core methodological elements remain essential. These include the clear definition of the functional unit, the establishment of appropriate system boundaries, and the consistent application of calculation principles such as allocation rules and cut-off criteria. Moreover, robust life cycle inventory models must be constructed for each system and design alternative under evaluation to ensure transparency, comparability, and methodological rigor.

The model was developed using GaBi 8.2, one of the leading life cycle assessment (LCA) software tools worldwide (thinkstep, Leinfelden-Echterdingen, Germany), together with the GaBi SP 34 database, which provides more than 10,000 environmental profiles as a robust basis for modelling [27]. The underlying framework identifies key biophysical systems that are threatened by human activities and establishes a quantitative structure for defining planetary boundaries that should not be exceeded in order to preserve the integrity of the global ecosphere. This concept offers a systematic approach for addressing the multiple pathways through which human actions contribute to environmental degradation and is continuously refined as new scientific insights emerge. Planetary boundaries have been adopted by several authors as a weighting framework within life cycle impact assessment [28]. Since the objective in this study is to derive a single aggregated indicator of environmental burden, the approach proposed by Sala et al. [9] is selected and adapted using the available distance-to-target normalization values. In parallel, both sink-related and source-related impact categories are applied to differentiate between areas of protection. Sink-related categories are directly associated with the biological systems they represent and can be collectively interpreted as indicators of global biophysical system stability [29]. Resource depletion is assigned to the category of global resource stocks, as the metabolic system described earlier depends fundamentally on the sustained availability of both biotic and abiotic resources. Although human-accessible resources are predominantly located underground and do not directly contribute to the stability of biophysical systems, they are nonetheless central to the broader concept of resource scarcity that is also evident in biological systems. To account for abiotic resource use, the anthropogenic stock extended abiotic depletion potential (AADP) model is employed, reflecting significant methodological advancements made since the original distance-to-target values were introduced. For land-use impacts, the biotic production indicator from LANCA 2.0 (Fraunhofer IBP, Stuttgart, Germany) is applied, with the updated and freely accessible version used in this assessment [30, 31]. Nevertheless, significant scope for improvement remains in assessing impacts on the global resource pool, especially in terms of incorporating temporal dynamics and regional differentiation. Within the adopted framework, the normalization factor derived from the planetary boundaries concept is applied to the results of each impact category across the entire life cycle. These normalized impact scores are then aggregated into a single indicator, which can be directly compared with the correspondingly derived single-point value of the reference systems.

7.3.2 *Environmental Function*

Within the proposed framework, functions are defined as the intended and expected properties of the system under evaluation. For environmental functions in particular, this requires a clearly articulated and demonstrable positive effect on the environment. A well-established biological analogy is mutualism, in which two species interact in a manner that is beneficial to both. Such biological exchanges can take several forms: service–service relationships, as seen in the mutual protection between sea anemones and anemone fishes; resource–resource relationships, exemplified by mycorrhizal associations between plant roots and fungi; or service–resource relationships, such as birds dispersing the seeds of fleshy fruits they consume. Beyond these direct interactions, nature is also characterized by numerous closed material cycles, including the carbon, nitrogen, sulfur, and phosphorus cycles, in which initial materials are periodically transformed and ultimately made available again through chemical and biological processes [26]. Most technical systems, however, are not designed to deliver a specific environmental function and are therefore not assessed under this criterion. Exceptions include systems that explicitly aim to improve the availability of their waste streams, such as waste treatment technologies, or those designed to convert wastes into forms that generate mutual benefits. In such cases, environmental functions can be explicitly identified, evaluated, and quantified. Positive expected impacts are assessed using the same methodological framework applied to environmental burdens, ensuring consistency in evaluation. By contrast, indirect life cycle–related effects, such as general recyclability, are not considered intentional design functions and are therefore accounted for within the burden assessment rather than as positive functional contributions [26].

7.4 Case Studies

According to Kong et al. [32], bio-inspired materials exhibit a wide range of mechanical, optical, and electrical functionalities. For example, Mei et al. developed honeycomb-inspired $CoMoO_x$ nanostructures composed of two-dimensional nanosheets with interconnected pores and channels. Compared with electrode materials of identical chemical composition but lacking a cellular nanostructure, these architectures showed enhanced structural robustness and improved lithium-storage capacity [33]. Electric eels (*Electrophorus electricus*) provide another striking biological model, as they generate electric current and high voltages through membranes containing densely packed ion channels that exploit ion gradients. Inspired by such mechanisms, as well as by structures found in hummingbird wings [34], human skin [35], and other natural surfaces, researchers have developed advanced triboelectric nanogenerators based on two-dimensional materials, along with ion-gradient power generators modeled after electric eels. In particular, eel-inspired polyacrylamide hydrogels have been shown to convert chemical energy into electrical energy, supporting the development of self-powered biomedical implants [36]. Further bio-inspired advances include the work of Ravi et al. [37], who reported electrical charge storage across multiple layers of photo-proteins isolated from *Rhodobacter sphaeroides*. These proteins demonstrate significant potential for light harvesting, charge storage, and the realization of self-charging biophotonic devices. Drawing inspiration from plant seed dispersal units, bilayer alumina compacts capable of folding via differential swelling have also been fabricated, enabling controlled shape transformation during sintering [38]. In the optical domain, Gur et al. [39] showed that the unique photonic structures of male sapphirinid copepods can inform the design of next-generation optical and light-harvesting devices. More broadly, studies indicate that biological materials derived from natural organisms often outperform conventional synthetic counterparts in applications such as electrocatalysis, photocatalysis, and biomass conversion, underscoring the functional advantages of bio-inspired design principles [40]. The application of biomimetic approaches remains constrained because conventional manufacturing techniques are often incapable of accurately reproducing the complex architectures found in biologically fabricated structures. One promising alternative is biofabricated self-shaping, a strategy for generating three-dimensional forms by modifying the internal structure of an initially flat material. When exposed to external stimuli such as moisture or temperature, these materials undergo controlled deformation, transforming into predefined three-dimensional configurations. Notably, bio-inspired self-shaping processes do not require electrical energy input and enable the creation of materials that autonomously respond to environmental cues [41]. Advances in three-dimensional (3D) printing technologies have significantly expanded the potential to fabricate more sophisticated bio-inspired structures. Growing demand for high-performance, function-specific materials has further accelerated the development of such techniques [42]. Owing to its inherent design flexibility, 3D printing allows the fabrication of structures with highly complex and arbitrary geometries that are difficult or impossible to achieve using traditional methods [43]. A range of additive manufacturing techniques—including direct ink writing, laser sintering, fused deposition modelling, multi-jet printing, and laser melting—have been employed to overcome existing constraints in producing biofabricated compositions [44]. Despite these advantages, 3D printing still faces notable challenges, particularly in terms of production speed and cost. Nevertheless, its versatility continues to attract substantial research interest. Applications include the fabrication of spinal cord–inspired scaffolds [45], ultralight biomimetic architectures based on cellular structures [46], and damage-tolerant construction units inspired by crystalline forms. Furthermore, the emergence of multi-nozzle 3D printing has enabled more precise control over material distribution, thereby substantially increasing the achievable diversity and functionality of printed bio-inspired materials.

7.5 Conclusion and Future Perspectives

It was found that integrating bio-inspired principles into sustainability concepts and sustainability assessment yields new insights and strengthens the connection to real-world development processes. Moreover, linking system characteristics derived from biological analogies was shown to improve practitioners' ability to understand and access overall sustainability performance. To demonstrate these outcomes, a simplified assessment approach is applied, in which each component such as functions and environmental burdens is represented by a single aggregated value. However, given the limited number of studies conducted to date, several fundamental questions remain unresolved. Continued application of the proposed assessment framework is therefore expected, particularly as the project within which it was developed progresses further. Future work will include a detailed examination of the relationship between resource demand and multifunctionality, considering both simultaneous and temporally separated functions. This analysis will treat functionality as a multidimensional set of functions and enabling properties that allow biological systems to persist while operating under resource constraints and environmental stress. The concept of bio-inspired sustainability provides a unified framework that combines a sustainability model with an assessment methodology encompassing all three pillars of sustainability, accounting for both intended benefits and unintended burdens. Its evaluation is performed using a consistent, life cycle based model that aims to deliberately guide the design of sustainable solutions while reintegrating principles from non-human biological systems into product development. Building on this initial implementation, the model will be further refined toward a more automated assessment tool and adapted for conventional, bio-inspired, and bio-based practical applications.

References

1. Katiyar NK, Goel G, Hawi S (2021) Nature-inspired materials: emerging trends and prospects. NPG Asia Mater 13:56
2. Zhang Y, Le Ferrand H (2022) Bio-inspired self-shaping clay composites for sustainable development. Biomimetics 7:13
3. Yu X-Q, Hu X-H, Zhu L, Li G, Guo M, Li Q, Xu C, Chen S (2021) Fabrication of magnetically driven photonic crystal fiber film via microfluidic blow-spinning towards dynamic biomimetic butterfly. Mater Lett 291:129450
4. Lee JC, Volpicelli EJ (2017) Bio-inspired collagen scaffolds in cranial bone regeneration: from bedside to bench. Adv Healthc Mater 6:1700232
5. Plaza A, Vargas-Silva G, Iriarte X, Ros J (2022) Mode-displacement method for structural dynamic analysis of bio-inspired structures: a palm-tree stem subject to wind effects. Wood Mater Sci Eng 17:1–15
6. Voet VSD, Kumar K, Brinke GT, Loos K (2015) Bio-inspired synthesis of well-ordered layered organic-inorganic Nanohybrids: mimicking the natural processing of nacre by mineralization of block copolymer templates. Macromol Rapid Commun 36:1756–1760
7. Pham M-S, Liu C, Todd I, Lertthanasarn J (2019) Damage-tolerant architected materials inspired by crystal microstructure. Nature 565:305–311
8. Speck O, Speck D, Horn R, Gantner J, Sedlbauer KP (2017) Biomimetic bio-inspired biomorph sustainable? An attempt to classify and clarify biology-derived technical developments. Bioinspir Biomim 12:11004
9. Sala S, Ciuffo B, Nijkamp P (2015) A systemic framework for sustainability assessment. Ecol Econ 119:314–325
10. Finkbeiner M, Schau EM, Lehmann A, Traverso M (2010) Towards life cycle sustainability assessment. Sustainability 2:3309–3322
11. Horn R, Gantner J, Widmer L, Sedlbauer KP, Speck O (2016) Bio-inspired sustainability assessment: a conceptual framework. In: Knippers J, Nickel KG, Speck T (eds) Biomimetic research for architecture and building construction: biological design and integrative structures, vol 8. Springer, Cham, pp 361–377
12. Folke C, Carpenter S, Elmqvist T, Gunderson L, Holling CS, Walker B (2002) Resilience and sustainable development: building adaptive capacity in a world of transformations. AMBIO J Hum Environ 31:437–440

13. Moore FC (2011) Toppling the tripod: sustainable development, constructive ambiguity and the environmental challenge. Cons J Sustain Dev 1:141–150
14. Zamagni A, Pesonen H-L, Swarr T (2013) From LCA to life cycle sustainability assessment: concept, practice and future directions. Int J Life Cycle Assess 18:1637–1641
15. Parent J, Cucuzzella C, Revéret J-P (2013) Revisiting the role of LCA and SLCA in the transition towards sustainable production and consumption. Int J Life Cycle Assess 18:1642–1652
16. Bjørn A, Hauschild MZ (2011) Cradle to Cradle and LCA—is there a conflict? Glocalized solutions for sustainability in manufacturing. In: Hesselbach J, Herrmann C (eds) Proceedings of the 18th CIRP International Conference on Life Cycle Engineering, Technische Universität Braunschweig, Braunschweig, Germany, 2–4 May 2011. Springer, Berlin/Heidelberg, pp 599–604
17. Onat NC, Kucukvar M, Tatari O (2014) Integrating triple bottom line input–output analysis into life cycle sustainability assessment framework: the case for US buildings. Int J Life Cycle Assess 19:1488–1505
18. Guinée J (2016) Life cycle sustainability assessment: what is it and what are its challenges? In: Clift R, Druckman A (eds) Taking stock of industrial ecology. Springer International, Cham, pp 45–68
19. Lai WL, Sharma S, Roy S et al (2022) Roadmap to sustainable plastic waste management: a focused study on recycling PET for triboelectric nanogenerator production in Singapore and India. Environ Sci Pollut Res 29:51234–51268. https://doi.org/10.1007/s11356-022-20854-2
20. Franco IP, Morales-Masis M, Mora-Seró I, Vidal R (2025) Comparative life cycle assessment of lead-free halide perovskite composites/polymer for piezoelectric energy harvesting. Sustain Energy Fuels 9(16):4375–4391. https://doi.org/10.1039/d5se00717h
21. Zhang H, Gan X, Yan Y et al (2024) A sustainable dual cross-linked cellulose hydrogel electrolyte for high-performance zinc-metal batteries. Nano Micro Lett 16:106. https://doi.org/10.1007/s40820-024-01329-0
22. Xie S, Yan H, Qi R (2024) A review of polymer-based environment-induced nanogenerators: power generation performance and polymer material manipulations. Polymers 16(4):555. https://doi.org/10.3390/polym16040555
23. Sharma S, Lai WL, Roy S et al (2024) Gate-to-grave assessment of plastic from recycling to manufacturing of TENG: a comparison between India and Singapore. Environ Sci Pollut Res 31:42698–42718. https://doi.org/10.1007/s11356-024-33867-w
24. Sureau S, Mazijn B, Garrido SR, Achten WMJ (2017) Social life-cycle assessment frameworks: a review of criteria and indicators proposed to assess social and socioeconomic impacts. Int J Life Cycle Assess 30:181
25. Jørgensen A (2013) Social LCA—a way ahead? Int J Life Cycle Assess 18:296–299
26. Horn R, Dahy H, Gantner J, Speck O, Leistner P (2018) Bio-inspired sustainability assessment for building product development—concept and case study. Sustainability 10(1):130. https://doi.org/10.3390/su10010130
27. Thinkstep AG (2017) GaBi TS: Sofware-system and databases for life cycle engineering. Thinkstep AG, Leinfelden-Echterdingen
28. Ryberg MW, Owsianiak M, Richardson K, Hauschild MZ (2016) Challenges in implementing a planetary boundaries based life-cycle impact assessment methodology. J Clean Prod 139:450–459
29. Sandin G, Peters GM, Svanström M (2015) Using the planetary boundaries framework for setting impact-reduction targets in LCA contexts. Int J Life Cycle Assess 20:1684–1700
30. Schneider L, Berger M, Finkbeiner M (2015) Abiotic resource depletion in LCA—background and update of the anthropogenic stock extended abiotic depletion potential (AADP) model. Int J Life Cycle Assess 20:709–721
31. Vidal Legaz B, Maia De Souza D, Teixeira R, Antón A, Putman B, Sala S (2017) Soil quality, properties, and functions in life cycle assessment: an evaluation of models. J Clean Prod 140:502–515

32. Kong T, Luo G, Zhao Y, Liu Z (2019) Bio-inspired superwettability micro/nanoarchitectures: fabrications and applications. Adv Funct Mater 29:1808012
33. Yang Y, Song X, Li X, Chen Z, Zhou C, Zhou Q, Chen Y (2018) Recent progress in biomimetic additive manufacturing technology: from materials to functional structures. Adv Mater 30:e1706539
34. Ahmed S, Brockgreitens J, Xu K, Abbas A (2017) A nanoselenium sponge for instantaneous mercury removal to undetectable levels. Adv Funct Mater 27:1606572
35. Wang L, Xie Y, Yang J, Zhu X, Hu Q, Li X, Liu Z (2017) Insight into mechanisms of fluoride removal from contaminated groundwater using lanthanum-modified bone waste. RSC Adv 7:54291–54305
36. Schroeder TBH, Guha A, Lamoureux A, VanRenterghem G, Sept D, Shtein M, Yang J, Mayer M (2017) An electric-eel-inspired soft power source from stacked hydrogels. Nature 552:214–218
37. Ravi SK, Rawding P, Elshahawy AM, Huang K, Sun W, Zhao F, Wang J, Jones MR, Tan SC (2019) Photosynthetic apparatus of Rhodobactersphaeroides exhibits prolonged charge storage. Nat Commun 10:902
38. Bargardi FL, Le Ferrand H, Libanori R, Studart AR (2016) Bio-inspired self-shaping ceramics. Nat Commun 7:13912
39. Gur D, Leshem B, Farstey V, Oron D, Addadi L, Weiner S (2016) Light-induced colour change in the Sapphirinid copepods: tunable photonic crystals. Adv Funct Mater 26:1393–1399
40. Gong C, Sun S, Zhang Y, Sun L, Su Z, Wu A, Wei G (2019) Hierarchical nanomaterials via biomolecular self-assembly and bioinspiration for energy and environmental applications. Nanoscale 11:4147–4182
41. Le Ferrand H, Studart AR, Arrieta AF (2019) Filtered mechanosensing using snapping composites with embedded mechanoelectrical transduction. ACS Nano 13:4752–4760
42. Li Y, Mao H, Hu P, Hermes M, Lim H, Yoon J, Luhar M, Chen Y, Wu W (2019) Bio-inspired functional surfaces enabled by multiscale stereolithography. Adv Mater Technol 4:1800638
43. Spadaccini CM (2019) Additive manufacturing and processing of architected materials. MRS Bull 44:782–788
44. Nazir A, Abate KM, Kumar A, Jeng JY (2019) A state-of-the-art review on types, design, optimization, and additive manufacturing of cellular structures. Int J Adv Manuf Technol 104:3489–3510
45. Koffler J, Zhu W, Qu X, Platoshyn O, Dulin JN, Brock J, Graham L, Lu P, Sakamoto J, Marsala M et al (2019) Biomimetic 3D-printed scaffolds for spinal cord injury repair. Nat Med 25:263–269
46. Peng M, Wen Z, Xie L, Cheng J, Jia Z, Shi D, Zeng H, Zhao B, Liang Z, Li T et al (2019) 3D printing of ultralight biomimetic hierarchical graphene materials with exceptional stiffness and resilience. Adv Mater 31:e1902930

Index

M. Y. Mir, J. A. Parray, *Nanoenergy Production*, SpringerBriefs in Energy,
https://doi.org/10.1007/978-3-032-20859-0

The manufacturer's authorised representative in the EU is Springer Nature Customer Service Centre GmbH, Europaplatz 3, 69115 Heidelberg, Germany. If you have any concerns regarding our products, please contact ProductSafety@springernature.com

Printed and bound by CPI Group (UK) Ltd, Croydon, CR0 4YY
07/07/2026
02160932-0001